Economics
of
Abundance

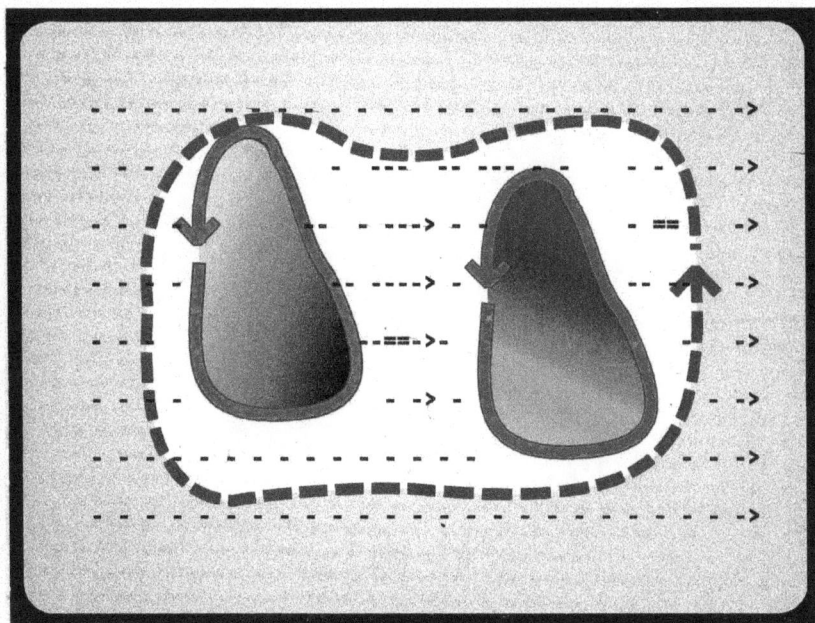

the Bob Komives
Conjectures

Economics of Abundance
The Bob Komives Conjectures

In Summary

In Summary

A. Derive Economics from Life Science.

The rules of wealth are written in the biosphere, by the biosphere. We are privileged to discover those rules slowly and to use them consciously. We can look beyond the apparent scarcity of the moment to the ubiquitous abundance that has nurtured evolution of our biosphere. Until the next cosmic disaster sets life back or extinguishes it, evolution builds ever more plentiful life from a once inanimate universe. Economics must better connect itself to life science to understand abundance.

Oil is dirty water
to anyone who has no knowledge of how to use it.
Water is death
to many who do not know how to swim or fish.
Fish may be angels or devils
to people who know no use for them.
Knowing that many fish are edible and tasty
has little value to a hungry quadriplegic
whose arms no longer know how to grasp a fish.
Knowing how to grasp a fish
has little value if it destroys all the fish—
eliminating the knowledge they carry within them.
                                    Knowledge Of Dirty Water

Our wealth is complex. It includes art, safety and clean air, as well as more tangible goods and services. A resource is a resource because life has knowledge to leverage it. Knowledge is wealth and has no inherent limitations except the limits of the universe. We must measure knowledge, track it, and understand how life organizes it if we are to understand abundance. The biosphere is rich in knowledge bound into energy forms that know how to organize themselves to convert more and more of the inanimate universe into life. Humankind's wealth is its share of that knowledge. Today, other life sciences know more about this wealth than does economics.

Economics usually deals only with humankind—as do some other sciences such as sociology. That limitation is legitimate since all sciences are really arbitrary sub-sciences of one whole, science. If intellectual boundaries are understood to be convenient rather than absolute, we let one science flow naturally into another. Economic analysis should work at species level, national level, regional level, group or family level, or at the level of any life system we choose. Why? Because wealth develops at all levels. Choosing any one level for analysis, we must look inward to sub-systems, but also outward to the systems of which the chosen level is a working part.

A community fosters its economic development when it captures wealth, distributes it, and recirculates it through community organs—neighborhoods, schools, families, businesses, associations. In the process of economic development new subgroups may form, and old ones may reorganize. As in nature, the successful process is at once complex and elegant.

This same community should look outward to try to understand the larger governments, cultures, ecosystems, and ultimately the biosphere and universe of which it is an organ. Economics, the life science, will tell us how, while capturing wealth from these higher-level organisms, we should foster their wealth—recognizing that they too develop and change. Economics, the life science, will explain to us why, if we organize ourselves to foul or stagnate our planet, we organize for failure.

B. Abolish Monetary Recall.

We try our dutiful best to pay taxes to the national government using money it has issued. Our government dutifully tries to collect this money as a tax. But we fail;

government fails. Rather than fulfill the duty of taxation, together we play the folly of monetary recall. Can anyone seriously believe that he or she would make the same gross income today if there were no national income taxes? For that is what you must believe to deny the folly. Our wages and salaries inflate to compensate for the taxes we must pay. We discount each dollar of earnings before we receive it. A new national tax will pull some money out of our pockets. It takes us a while to adjust. But, we bargain; we quit; we strike; we bargain. We force the economy to inflate itself enough to give us back the money we lost. The net income to the federal government from monetary recall?—nil.

We do not need to recall money to pay for our national government. However, since we now receive inflated incomes due to these pseudo taxes, If we were to eliminate the taxes overnight we would likely put dollars into our hands faster than we could invest them wisely. That would cause inflation. Let the national, recall-based tax system die a slow but steady death. National governments should retain their power to impose surprise recalls of money in case a series of badly invested national budgets causes an inflation emergency. However, these non-revenue recalls should be instant, one-time, and quickly burned.

C. Abolish National Borrowing.

This is all unfortunate, but necessary, says the economist in his scarcity model of the world.
We cannot provide for the welfare of the many
because resources available to the few are scarce.
We cannot clean up and protect our scarce resources because,
again, as the logic goes,
our resources are scarce.

Where is the evidence
that this scarcity model works?

Where is the evidence
that doing the right thing costs too much,
that doing the wrong thing is affordable?
Where is the evidence
that poor countries
that follow this model
will someday become rich?

Where is the evidence?

Evidence

Monetary recall is not needed to pay for national government. National borrowing is a substitute for these pseudo taxes. Therefore, there is no need to borrow. When a country borrows foreign money from a foreign government, it shackles its own economy with unneeded obligations. When a central bank fights inflation by raising interest rates, it fights a fire with kerosene. A national government cannot enter the borrowing business, whether internal or international, without raising inflationary pressure. While local banks play a key role in financing local economic development through loans, a central bank such as the Federal Reserve System in the USA need have no role in financing national expenditures. Once this is understood, the central bank may assist a healthy flow of money into the economy through independent local banks.

D. Print Money and Invest Wisely

Today, we print our money on valuable paper. Corporations print common stock on cheaper paper. Banks print unsecured loans on the most ordinary of paper. Stocks and loans are forms of private money. They become worthless more easily than their sibling, national currency. No one seems to complain that they are paper.

What works for banks and corporations works for our national governments. Our politicians can print money to pay for good projects. They should never print money—or tax it or borrow it—to pay for bad projects. The key does not lie in the printing but rather in the investment. A good project meets moral criteria, retains the value of the resources invested, complements the biosphere, and creates new wealth that equals or exceeds the face value of the money spent.

Projects that meet these criteria in an economic sense but fall somewhat short as measured in the marketplace may still be good projects. The low inflation that they cause has proven to be tolerable.

Politics can work for us. Our leaders can debate the investment potential of each law and expenditure. If we know how government economics really work, we may teach our officials to manage our future responsibly. We will debate as hotly as ever, but our disagreements can be more productive. No resources will be wasted on false taxes, false borrowing from citizens, or useless borrowing of foreign currency from foreign nations. We will debate projects and policies on their broadly defined economic value. Within the limits of our expanding knowledge of economics and our insight to the future we will fund projects on their merits. We will subsidize meritorious private efforts explicitly, not through loopholes hidden in a useless tax code.

Economic development strategies and our debates about these strategies will improve as economics integrates other life sciences. We will reach out more to the energy of the sun and the solar system—to increase biospheric capture rather than destroy biospheric wealth. We will recirculate energy productively through natural and man-made systems. As we learn, we will do more and more with less and less and (if we wish) do it faster.

If we invest our money in complex and just distribution of wealth among people of varied skills and tastes, we will deepen our wealth, keeping captured energy at the service of our species for an ever increasing time. Our byproducts will be less waste and more product as we recirculate them back through our distribution system. Our wealth and the biosphere's wealth will grow.

E. Recycle Misguided Intelligentsia.

Restless need for knowledge precedes development of science. There will always be voids in knowledge, unanswered questions. The more restless the people who feel a need to fill the voids, the more energy that society diverts into science. If acceptable answers do not come forth, we may divert energy from science into its sexier counterpart, pseudoscience. The struggling scientist who knows very little can feel unneeded in the shadow of pseudoscientists who profess wisdom. Do I profess pseudoscience? I am not wise enough to know. I hope to shed light and doubt I will cast a shadow. Allow me a flash of self pity as I smile at either prospect. Woe to me if I ever cast a pseudoscientific shadow.

We have educated some of our best people to do nothing of value. A significant portion of U.S. America's intelligentsia collects, calculates, or works to avoid federal recall of federal money. Many speculate on the affects of federal taxes and loans. Our leadership intelligentsia wastes much of its time debating how best to undertake the useless loans and monetary recall that are supposed to balance our federal budget and guide the economy.

This is high alchemy. Only smart people qualify. These sincere, talented people certainly outnumber our declared welfare recipients. Yet, despite being functionally unemployed and consuming large quantities of our resources, they often receive both respect and high pay. Imagine the boost to our economy if the efforts of this elite were recycled into beneficial work.

F. For Peace, Use Abundance.

War is age-old medicine
for peacetime economic myth.
It can spur us to do the unconventional.
If willing to abandon dreams of community,
we can again abandon fiscal conservatism

in order to wage war,
but the pay-offs and risks have changed.
High-tech, "conventional" war is deadly,
but, after all,
just conventional.
Whole-wax, nuclear war is more deadly,
but, after all,
just unthinkable.
If we will refuse to use war
to cure ills in the economics of scarcity,
we must learn to use abundance
to cure ills in the economics of peace.

‖

G. Let It Comfort Humankind.

My version of economics will not make everything right. No version, no matter how good, will be permanent. We are cursed and blessed by an imperfect evolution. Yet, I believe that if we build a science of economics on principles of abundance we will participate more productively in our evolution. From a viewpoint prejudiced by my species, I say that human evolution has been more good than bad. I say this comfortably, but not too comfortably.

We learn from failure, successes.
We also unlearn.

Dinosaurs:
   millions of years,
   billions of successes,
   a grand experiment,
   high-class guest,
   then biosphere forgot how to make them.
Humankind:
   new guest,
   still learning,
   biosphere has just begun our experiment.

Among those
species,
cultures,
families,
individuals
who survived yesterday,
many,
but not all,
are fit.
Among those that did not survive,
many,
but not all,
are unfit.
Who will survive until tomorrow?

Countless blunders in the biosphere today.
Some who blundered,
some who innocently stood by,
caught by biospheric justice
—executed before tomorrow.

Let it scare humankind;
again our capacity to blunder rose today.
Let it comfort humankind;
again rose our capacity to understand,
   anticipate,
   record,
   improve
what survives today until tomorrow.

Experiment Just Begun

Into this comforting rise of capacity I publish this work.

Part One:
Life's Inventive, Imperfect Invasion
— topics 1 - 35—

1. A Faint Road

I found a faint road through a vast field
where genius, fool, and charlatan must ply.
As hard as the road is to follow,
harder still is to know who am I.

On The Art Of Synthesis

I am a wealthy pattern in my young, abundant biosphere. I am a thread in the net of life that threatens to encircle the universe. I seek a science to incorporate both the elusive abundance that builds what I have and the apparent scarcity that every day shows me what I have not. Through a vast field I follow a faint road along which I see landscapes that are impenetrable to traditional machinery of national and international finance. I see a distant village of mainstream economics barred from these exciting landscapes by its own walls and by the militant forces of pseudo-economics that interpose quaint, mirage-landscapes for mainstream society to fancy. In the same light that bathes the backs of those who once argued for a flat earth, I see proud, hoping, and helpless faces of those who argue for this week's popular economics— balance-the-national-budget-or-die. I see victim and perpetrator of quaint fancy.

Perhaps one fancy can replace another. Perhaps I can point through patches of scarcity in a field of abundance to a faint road that you will fancy to explore. The expanded substance of Economics of Abundance came with Plum Local III (1991). I revised its organization in 1998 with Plum Local IV: an Essay on the Economics of Abundance. Now, with many small tweaks, I present what I hope is the final edition. For better and worse, what I offer you here has changed little over the decades..

11

2. Everybody Knows

In 1990 and 1991, when the United States of America lead
other countries in a war to evict the forces of Iraq from
Kuwait, many advocates of balanced national budgets knew
it was time to abandon that principle in order to wage war. I
heard nobody ask, "If debt for war is good, can debt for
peace and public welfare be bad?"

A war rages in the Middle East
—costly by measures more important than money.
We so readily
suspend our fantasy of a balanced budget
so that we may fight a harsh war,
only to again impose our fantasy,
with harsh futility,
during brief interludes of peace.

Oh, the insidious fantasy!
Never apologize for expenditures
if they do not exceed taxes.
That is, if government recalls from us
at least as much money as it spends,
it can boast:
" We ruined the country
and much of the rest of the biosphere,
but we never ran an unbalanced budget."
Let us remember
that war is the age-old medicine
to counter peacetime fantasies.
For failing to make good investments in peace
we are as likely as ever
to
fall
into
internal and external
conflict
that will lead us again to war.

from: A War Rages

Things are not as they should be. The cold-war dichotomy between communism and capitalism has blurred. It should now be easier to study the complementary relationships between socialism and marketplace, and between peace and investment. Yet, in the years since the war in Kuwait such discussions seem less frequent, or, at least, less noticed.

Also in 1991, leaders in the United States of America were in a panic over their failing banks. Those who had long advocated smaller, decentralized government were sure it was time for larger, more centralized banks. They now have them. I see irony in this past and problems in this future.

One panic replaces another. In 1996 everybody knew that the big problem in the USA was budget balancing—provided we increase military expenditures and decrease both our taxes on the wealthy and our assistance to the poor. In early 1998 the problem seemed to be what to do with a projected budget surplus if we do not wage war with Iraq. Yesterday and today everybody knows that, when convenient, national taxes must balance expenses.

Once upon a time, everybody knew
the earth is flat.
Common sense confirmed it.
Common politicians ratified it.
The best scientists of the day spoke doubts.
Since everybody knew,
nobody listened.

from: Everybody Knew

3. Plum Local

I wrote the core conjectures of *Economics of Abundance* in 1980 and 1981 with *Plum Local* and *the 2ⁿᵈ Plum Local*. The expanded substance of this present version is from 1991's *for Love of Wealth & Biosphere*. The present format, with 92 concise topics, appears in the 1998 *Plum Local IV, a Primer on Economics*. I give only minor tweaks and new title to that primer to publish here, in 2019: *Economics of Abundance: the Bob Komives Conjectures.*

I had my first course in economics in college. I forgot most of it, except for the fascinating way that banks create money as they lend out most of the money that we deposit with them, then receive most of it back again in new deposits, and then lend most of this magically expanding cash out again, and on, and on. We see that bad banks fail, and we know that even good banks make bad loans.

Why only blame
—if our banks create money—
why only blame our government for inflation?

‖

I turned my studies to art and architecture. Along the way I discovered a maverick named R. Buckminster Fuller. He stood among other heroes such as Louis Sullivan and Frank Lloyd Wright in describing the unity of design and nature.

Can projects designed
following principles of our biosphere
ever be too ugly,
ever be too expensive?

‖

I got married, and we went to the Peace Corps near the Pacific Coast in Guatemala. I saw discrepancy between strategies for national economic development and realities of community development.

I had to ask
" Does it make sense
—for our poverty, our sickness, our exploitation—
that our cure
cannot come
with our economic development,
but only after?"

‖

We moved to Little Rock Arkansas where I tried my hand at city planning in the Model Cities program. This was 1969, a time of large investment in troubled cities. Our successes were real but modest.

Is it not strange?
Even during prosperous times
since our era of generosity,
they say,
we cannot afford to budget for success.

Since Our Era Of Generosity

I went back to school to get my professional planning degree. There I discovered economics, learning its many applications to local public policy. It was elegant; it was beautiful. The curves conveyed information to me in ways that no other medium ever had.

One weekend, I took a rest from my studies and read a book by R. Buckminster Fuller. I believe it was Operating Manual for Spaceship Earth. There, I encountered for the first time his elegant formulation of the fundamental law of economics:

Wealth is a function of energy and knowledge.

Absent was any mention of scarcity, supply, demand. This was the economics of abundance. Fuller's economics made every bit as much sense to me as the crisp logic of market economics.

Humankind developed
laws,
traditions,
and institutions
to deal with scarcity.
At any point
in time and space,
scarcity is specific.
It is real.

We live scarcity,
but we come to live

15

and to thrive
through abundance.

Please do not misunderstand me.
I believe in scarcity.
I have lived it and seen
both its pains
and its benefits.

Yet, abundance is as real as is scarcity
and is even more fundamental.

Without scarcity,
the economist cannot draw
supply curves
and demand curves.
But these curves cannot anticipate
mathematics,
art,
democracy,
communities,
back rubs,
interplanetary exploration,
civil rights,
the popsicle,
or the yo-yo.
Nor could they have anticipated
the brown trout,
the monarch butterfly,
or the horned toad.
Each is part of our biosphere.
Each is our wealth.
And wealth must be the stuff of economics.

Today,
if we choose to love our wealth and our biosphere
we seem unable to seek the best for one
without harming the other.

Today, also,
sages preach to us of the evils in our economy.
They tell us to be more moral,
to separate pretension from wealth.
Let us heed such sermons.

Yet, the moral sage does not free us
from the choice between two loves.
Neither sage nor economist can free us
unless we know
how wealth and biosphere are one—
how we live scarcity,
but come to live
and to thrive
through abundance.

We Come to Live and Thrive

I lay sandwiched between a straightforward explanation of
supply-demand-utility and Fuller's statement that wealth is a
function of knowledge and energy. I found myself in that
muddled layer of confusion and witchcraft called
macroeconomics—including gold flow, balance of payments,
balance of trade, inflation, and the like. The economy uses
the biosphere's model of abundance, while conventional
economics uses a model of scarcity. Beneath scarcity lies a
supportive abundance—a macro-abundance. Beneath
microeconomics, which specializes in scarcity, should lie a
supportive macroeconomics specializing in abundance.

Microeconomics covers those situations in which flow of
wealth mimics a traditional marketplace. People buy; they
sell; they trade. The demand for a product in relation to its
supply sets the price. Economists do not. Buyers and sellers
do so, acting upon their needs and desires. One day, two
chickens are worth two yards of cloth. The next day, they
may be worth three yards in the morning but only one after
lunch. Marketplace economics explains well the dynamics in
this true marketplace and in myriad public and private
markets in which goods and services are bought and sold. It
can explain how the price for cloth changes as well as how

the weaver decides how much to produce. It cannot, however, go on to explain how cloth came into existence nor how chickens were domesticated.

From graduate school I launched my planning career. I went to the island of Martha's Vineyard where I worked for five years to protect its resources and foster sensitive development. I moved as a consultant to Colorado, worked for a while in the analysis of socioeconomic impact from energy development. I went on to typical land-use planning. The gulf between my professional work and my struggle with the theories of economics seemed unnecessary, but enormous.

The economics of abundance remained a closet hobby until 1980 when I pulled together some of my notes in a hand-printed, ten-page document called *Plum Local* (see Appendix). It began with my apology: "Pardon my boldness ." I wrote, "The valid world economics will show the tie between genetic and economic evolution," and "Taxes, Bah!! Let's phase them out Let's balance our budget by investing communally (politically) in the growth of knowledge for mankind." I sent one copy to R. Buckminster Fuller. When I received his encouraging one-sentence response I felt some comfort.

The four-page *2nd Plum Local* of 1981 took my ideas further: "Taxation is role playing. Monetary return as we have in the national income tax system has no role to play. If there is a utopia it will be found in a humanistic management of instability."

Now, as then, I find it hard to put forth theories of economics that disagree with the teachings and preachings of intelligent people who are economists by profession. However, I would find it harder not to share ideas that help me find some sense and science among a potpourri of confusing theories and popular maxims.

4. A Crazy Cycle

In 1984 I had a chance to return with my family to Central America to teach land-use planning and work in watershed management. There, the importance hit me of bridging the gap between my land-use-planning work and my hobby of cogitating economics.

It made no sense to me.
The International Monetary Fund said
Central American governments should
reduce investment in
social
health
and environmental programs
(that were slowly raising quality of life for their citizens)
in order to borrow money
to pay for projects
that would produce exports,
in order to bring in outside money
to spawn development
to trickle down some resources,
in order to bring back
social
health
and environmental programs
to slowly raise the quality of life for their citizens.

A Crazy Cycle

5. Economics Should be Life Science

In life there is everywhere synergism. Two or more organs, two or more organisms, act together to achieve what neither could achieve alone. The whole is greater than the sum of its parts. R. Buckminster Fuller called it "synergy."

Scarcities are microeconomic parts
within macroeconomic abundance.

True to synergy,
we cannot divine the behavior of our abundance
if we study only our scarcities.

‖

Economists have failed to build a unified theory of abundance out of their keen understanding of scarcity. Why? It cannot be done. Many scientists work to overcome this problem under the general umbrella of the study of complexity. The Santa Fe Institute in Santa Fe, New Mexico is one focal point for their work. Such terms as artificial life, self-organizing economies, and increasing returns cover specific topics that draw away from the traditional static models of economics toward the dynamics that we see in nature—that we see in the economy but cannot explain using standard models. I feel confident that their work will deal with both abundance and scarcity and will eventually revolutionize economics–beyond even their expectations.

Economics must be a science
—a human science,
a life science.
Wealth is basic.
It cannot be the invention of humankind.
It must be traced to the bases of life.
It cannot reside mainly in banks and buildings,
moneys and stock.
Primitive people had wealth
but none of these.

‖

Science is a whole. It tries to encompass all that exists as well as all that may have existed and all that may come to exist. It attempts to describe the rules that govern the universe at all scales and make useful predictions based on these rules. The bodies of theory and information that we call the sciences are artificial, but convenient subdivisions of the whole, science. Physics cannot be separated from chemistry nor from sociology. At their frontiers, and even at their centers, there is overlap and synergy.

6. *Some Good Hands*

Economics may be the academic field that has paid least heed to the unity underlying the several sciences. It offers few if any connections to the vast underpinnings of science. We do call economics a science but see little clue where to fit its complex maxims, axioms and curves. It tries to describe reality but often does so in arcane ways that more separate it than ally it to the rest of science. This will change.

Serious and capable scientists, serious and capable economists, work to rectify this problem. I mention a few. Paul Krugman argues well that good-old-fashioned Keynesian economics is much better than the fads that have dominated public policy discussion in recent decades, but he also works at a cutting-edge economics-evolution. I heard him say that economics and evolution are almost the same subject. Paul Romer has brought technology and growth back into the mainstream of economic discussion. He notes that economists tell us nothing about why economic growth occurs (Economist, Feb. 5, 1996). At the Santa Fe Institute, Brian Arthur and others representing several fields of science have applied their growing understanding of complexity from the "hard" sciences to economic phenomena. They collaborate with colleagues from around the world (including Krugman and Romer), and they build from the insights of earlier scientists whose work was bypassed by the mainstream of economic thought. Edward O. Wilson calls for Consilience, a "jumping together", among the natural and social sciences and the humanities. He argues well why economics must incoporate the natural sciences.

I believe the impact of such work and thought will be more revolutionary than even its enthusiastic supporters now project. I think they will eventually come to conclusions similar to mine, though I doubt any of them could now agree. My arguments will not change their minds; their own research must do so. I do believe that most scientists who work at the cutting edges of economics could agree that mainstream economics has long suffered from poor connections to the vast underpinnings of science.

7. Science and Pseudoscience

While we can compare theory and experiments in microeconomics with findings by historians, anthropologists, sociologists, game theorists, and psychologists; microeconomics provides few connections to the rest of science. It has a fair excuse; it makes no pretensions. Microeconomics limits itself to a narrow subject, the marketplace and interactions that resemble the marketplace.

It is macroeconomics that should connect all of economics to the rest of science. After all, it should take a macro-view, look at the whole. It should overlap in many places with the other sciences. Narrow-viewed microeconomics should nestle comfortably inside. I see the reverse to be more true. Microeconomics forms the underpinnings of today's Macroeconomics. An unfathomable web of rules and rationalizations has spun out of marketplace theories of supply and demand to bind together the larger world of macroeconomics.

For macroeconomics we have something more akin to pseudoscience. It seems to patch together theories that rationalize all of yesterday, fail to predict tomorrow, and that do not lend themselves to testing . The young science of complexity shows that we cannot always expect to predict tomorrow; so prediction cannot be the only test of science. Simple parts can synergize a future whole that has recognizable but unpredictable patterns. There is a better test of science: falsifiability.

Science and pseudoscience are incompatible: astronomy and astrology, evolution and creationism. The separation of macroeconomics from science comes from mutual repulsion between two inherently incompatible bodies of thought. I wish to paraphrase and thank a scholar who commented on the difference between creationism and evolution. I did not hear his name when a radio network interviewed him in 1987 during one resurgence of controversy over those competing explanations of the origin of our species. The difference? Evolution could conceivably be disproved by evidence whereas creationism could not. Thanks to Murray Gell-Mann and his book, The Quark and the Jaguar, I now know to give some credit to philosopher Karl Popper who promoted this falsifiability test for science.

Scientists modify the theories of evolution as they gather new evidence. Creationists are bound by their belief to support their story of creation no matter the changes in the evidence. Today's descriptions of evolution may not be perfectly correct, but they are science. Creationism is a complex pseudoscience that mounts evidence to defend a belief.

Good macroeconomists do practice science. They subscribe to the principle that their theories could be falsified by evidence. But macroeconomics has become so abstruse—and at times defensive—that it resembles a belief system to be manipulated by politicians rather than nurtured by scientists. Depending on your political viewpoint and the latest fad, macroeconomics is a particular dogma. In 1998 "everybody knows" the federal budget must balance. For a politician in the United States of America to voice doubt in balanced federal budgets might be as harmful to her political career as expressing doubt in the bible. This seems to be the political reality, though even mainstream economics taught every day in our universities gives little importance to budget balancing.

Part of the problem here is that much of the economics that sets public policy has been wrested from the hands of economists. The media seem every day to find someone who calls himself "an economist for the Wall-Street firm of Stock, Broke, and Bond" to say that the market went up, or down, or failed to do either, "because president and congress failed to reach an agreement today to balance the budget (or distribute the surplus)." Of course, even if one statement happens that day to be true, fluctuations in today's market do not prove that balanced budgets are necessary any more than buying an umbrella proves that umbrellas cause rain. The public, however, is left to conclude that what it already knows to be true is true: a federal balanced budget is necessary.

> *Once upon a time, everybody knew*
> *the earth is the center of the universe.*
> *This was confirmed by religion*
> *and ratified by politicians.*
> *The best scientists of the day spoke doubts.*
>
> *from: Everybody Knew*

The public of 1633 in Europe must also have concluded that what it knew to be true was, in fact, true. Its media reported correctly that Galileo had just recanted his published conclusion that Copernicus was correct. Standing before judges of the Inquisition, Galileo said, no, he no longer believed what he and Copernicus had written. No, the sun is not the center of a solar system of planets. Yes, the planets do revolve around the earth, the center of the universe.

> *You think Galileo was a great one,*
> *but he wrote heresy in 1632.*
> *He wrote, " Copernicus is right,*
> *our earth circles the sun,*
> *not the other way around. "*
> *You think Galileo was a great one,*
> *but in 1633 he did recant so he would not burn.*
>
> *Now you too believe*
> *that to the sun belong the planets,*
> *that we live on one example of them,*
> *our sun-centered revolution,*
> *a scientific revelation,*
> *from a genius then among them*

a religious revolution,
insult to god above them.

One way or another,
believers go early, but truth stays late.
Yes, die for your country to get a plaque.
Yes, die for your religion to get guaranteed heaven.
But why die for your science to get guaranteed hell?
Why should you burn for your solar system?

Is the martyr more hero than the genius?
We well know how to make you a martyr,
but we lack the weapon to make you a genius.

How can you resist?
How can you insist:
that Earth is a sphere,
if it is healthier to talk "flat"?
that we came from evolution,
if the inquisition favors special creation?
that all peoples are equal,
if we preach one-ethnic perfection?

Recant today so you will not burn.
Choose humbly to not-believe what you believe.
Humility is a sign of greatness.
For everybody knows and the bible humbly shows
our Earth to be center to the universe.

You think Galileo was a great one
for finding the motion of the pendulum,
the equal rates of falling objects,
and, of course, our telescope.
But then he wrote that Copernicus is right.
You think Galileo was a great one
—but then he did recant
—but then he did not burn.

You Think Galileo Was A Great One

What-everybody-knows is often not true. What everybody knows about the economy is often wrong, but serious economists, when they are asked, seem unable to help the public know better. For example, while they can tolerate unbalanced budgets, we hear serious, academic economists stand before a media inquisition and say about national debt:

"We know at some point too much is too much."

"We do not know how much is too much too much."

Such statements are as unclear as they are unfalsifiable. By comparison, those articulate, Wall-Street economists sound clear and confident. They give short answers that sound precise—even if, when spliced together, their daily pronouncements make neither sense nor science.

In search of sense,
in search of a science
I took my thoughts on a trip from economics
haphazardly
to the origins of life
and, still in search,
back again.

‖

8. Market and Vision Must Change

I suffer
when I see North American cities sprawl
inefficiently, sloppily across the landscape.
My heart hurts for the poor campesino
struggling to produce
a poor crop
on poor land
cut and burned
from Earth's diminishing forest.
Harsh reality,
or bad vision?

‖

When such problems occur in a market economy, the harsh marketplace seems to fail the biosphere and its human population. In socialist economies, it seems to be the planners and administrators who fail in pursuit of their vision of economic development. I say both market and vision must change.

9. Ignorance and Economic Development

Despite our ignorance,
we are fair instruments of economic development,
while often not fair to the biosphere which gave us life.
Remove more of our ignorance
and we may become consistent economic developers.

‖

I see economists shackled to a model of the world based on scarcity—making it difficult for them to approach the plentiful world of economic development. The last time I checked in detail, this was quite evident in text books. For example, Paul A. Samuelson and William D. Nordhaus, in the twelfth edition of Economics, devoted section seven (of seven) to economic growth and international trade, chapter 36 (of 40) to the theory and evidence of economic growth, and only 3.5 pages out of about 900 to "The Sources of Economic Growth." In these pages, they described growth accounting as an attempt to measure the ingredients that contributed to past growth trends—for example, capital, labor, land, education, and technological advancement. They stated frankly that no theory seems to fit reality very well. As for growth accounting, they wrote that it is far from perfect, but it is about as good a guide as any in this imperfect world. I felt that to be a discouraging conclusion to find on page 799 of an introduction to economics.

More encouraging were theories relegated to the appendix. There I found Joseph Schumpeter's model emphasizing innovation, Harrod & Domar's emphasizing productivity, and

Von Neumann's emphasizing a logical tie between the growth rate and the interest rate. These models apparently got relegated to the appendix because there is no unifying theory. Samuelson and Nordhaus expressed hope for a future synthesis that will integrate the neoclassical analysis of economic growth with some 300 pages of macroeconomic problems found earlier in their book.

Can economics be a mature science if its theories of economic growth do not mesh with what economists call macroeconomics? Other authors of other texts may organize things differently; I doubt that they improve much on the clarity and honesty of Samuelson and Nordhaus. The problem lies in economics, not Economics.

10. Buckminster Fuller's Elegant Law

Growth studies are integral to physiology. Construction is integral to architecture. Economic development should be integral to economics. Yet, economic developers and economists share little common ground. At least one economist, Paul Romer, works to change that by convincing other economists that knowledge and technology are part of economic growth. This seems obvious to a person on the street but revolutionary to most macroeconomists.

Depending on where I run into it, economic development still seems to be more in the kit bag of real estate developers, promoters, and financiers, as well as thought provoking generalists. Confusion among economists makes me no more comfortable with the developers and promoters who play on a simpler stage:

" I create jobs.
  You want jobs.
  Jobs make community rich.
  Give me break.
  I will make money.

Then, I will make you
a nice, rich community."

‖

Pending some coherent understanding of economic development such a monologue plays well to many audiences.

I find more comfort with the generalist synthesizers who try to comprehend the complex world and take us along for the trial. They look beyond their narrow window on the universe to an ever-expansive view. This is not frivolous pursuit leading to nowhere. Some, such as da Vinci and Einstein left obvious legacies. Others do no less than help us think, create, and pursue our own synthesis. They are skilled people who are generalists and designers.

Generalists:
see pattern
where others see spots,
see hypothesis
where others struggle to see the experiment.
Designers:
see opportunities
where others see problems.

‖

One generalist designer got me onto this economic
tangent. R. Buckminster Fuller formulated the
fundamental law of macroeconomics. Wealth is a function
of the knowledge and the energy in the universe.

wealth = function of (knowledge & energy)

The amount of energy in the universe is constant. The amount of that energy that we can call wealth, however, is not constant. It varies as the biosphere grows and changes. According to Fuller's Law, then, knowledge is the only component of wealth that can cause wealth to grow or shrink. This makes sense if we adopt a broad concept of knowledge.

We know something about
food, clothing, shelter,
art, and human rights.
They know some things about us.
Each and all is wealth.
Food is energy
made palatable.
Clothing and shelter are energy
reformed into protection and image.
Art controls and molds energy
to please our senses.
Human rights let us control our own energy
(including our own bodies).
Lacking such energy,
and the knowledge to use it,
we are poor.

‖

Buckminster Fuller's formulation for wealth lead him to make a simple prescription for economic development: Increase knowledge! Since the quantity of energy in the universe is fixed, we create wealth only by increasing knowledge. More important, when we increase our knowledge we cannot fail to increase our wealth.

My own formulation is incidentally different, but it leads to the same conclusions:

wealth = knowledge

On this alone I would like to rest my case. What else I have to say is not so clean and simple. I too need more designer synthesis. However, I believe I can argue well that economic development is inseparable from economics. I can lay down some of the bases for a conscious economic development strategy that equates with a sane use of the biosphere and discards the popular conception of a balanced national budget. The knowledge we need is within us, around us, and ahead of us.

11. Classes of Knowledge, Uses of Wealth

Knowledge is just an abstract concept
until we tie it to the energy that it harnesses.
The wealth potential of the biosphere
is to have useful knowledge
of all the energy in the universe.
We have reached but the tiniest fraction of that potential,
but take note:
　　for billions of years the total was zero.
Once The Total Was Zero

Knowledge makes wealth. Wealth is energy—that energy that is of value because we know how to use it to our benefit. Even when ignorant in our brains of value, something in our body, our society, or our environment knows how to harness the energy of wealth to our benefit. One can benefit from electricity knowing little of its physics and mechanics. Nor is it necessary to know how our immune systems work to benefit from their protection.

As our knowledge increases, our species' wealth increases. As long as our knowledge increases faster than our population, wealth per person increases. Buckminster Fuller saw that if we understand the principles of knowledge formation we understand the principles of economic development. With such understanding we might avoid extinction and continue to partake in evolution's gift of growing prosperity.

Fuller saw energy as physical, yet saw knowledge as metaphysical. Since knowledge is the only variable in his conception of wealth, if it is metaphysical it does not lend itself to measurement. (At least I do not know any metaphysical measurements to apply.) Convinced that Fuller's Law held much more promise than the collection of theories that were passing for macroeconomics, I looked for a way to reformulate the law so that we might test it alongside other physical laws that we reveal through science. There should be notation that can express Fuller's Law in physical terms. Knowledge could then be discussed as a quantity even if measurement were still impractical.

After an enjoyable, but long, struggle with this measurement problem I saw that it simplifies if we think of one instant in time and ignore any complications from the relativity of time across the vastness of the universe. At one instant as seen from one point in the universe, knowledge is not metaphysical. It acts upon discrete units of energy. At that instant all of the energy of the universe can be divided into three classes.

Class-w energy is the knowledge (wealth) of a particular species, ecosystem, community, or other subordinate part of the biosphere. Human wealth is **class-w**human.

Class-W energy is biospheric knowledge. It includes all **class-w** energy.

Class-U energy is all the energy in the universe, including **class-W**.

Since knowledge and energy are inseparable at this instant, we can measure knowledge by measuring the energy to which it has attached. Our human wealth equation becomes a summation of this attached energy:

$$\sum \textbf{wealth} = \sum \textbf{knowledge} = \sum \textbf{class-w}_{\textbf{human}} \textbf{ energy}$$

When life appeared the amount of the energy in the universe devoted to life went from zero to a tiny amount. As life multiplies that amount stays tiny relative to the universe but steadily increases. Life's invasion of the inanimate universe has begun. Earth has become a biosphere. Life captures inanimate energy and organizes it into its systems. Some newly captured universal energy gets bound into living creatures. A larger amount—still inanimate—serves as storehouse of energy for life support. Life exploits this storehouse as food and as supportive environment in which to survive and prosper. This captured, organized energy is life's wealth. It includes the growing amount of energy that actually knows how to live and the expanding reservoir of inanimate energy that life knows how to exploit. So it is for our species. Our class-whuman energy is the energy that knows how to be we, plus other biospheric energy that we know how to use to our benefit.

That part of Class-Wbiospheric energy that we cannot count in our class-whuman wealth is quite important. It measures our ignorance.

12. Our Challenge: Grow into the Biosphere

*When we admit that parts of the biosphere
lie beyond our wealth estate
we receive no license to destroy,
but, rather, a challenge
—rather our challenge to integrate.*

‖

Exploration can lead down unexpected and unsought paths. I found it helpful to divide wealth into classes. However when I distinguished class-W (biospheric wealth) from our class-w (human wealth), I worried that I was about to conclude that we could do without much of the biosphere—that we can afford to destroy or abuse any of the biosphere that is not classified as human wealth.

For *Plum Local III* I sketched a series of diagrams that helped me to another conclusion.

Diagram 1

Once there was an inanimate universe.

Diagram 2

Life appears. The biosphere begins.

Diagram 3

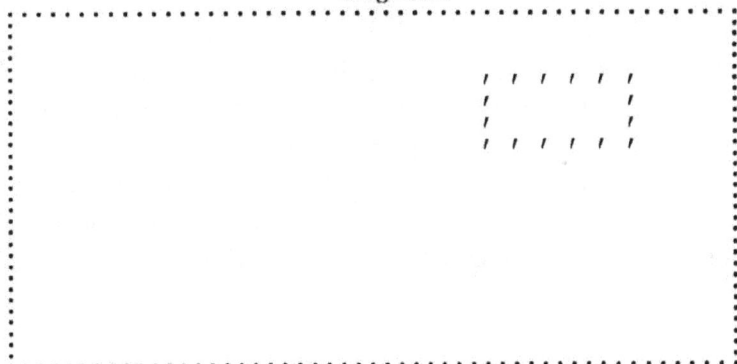

Life prospers as biosphere invades the inanimate universe.

Diagram 4

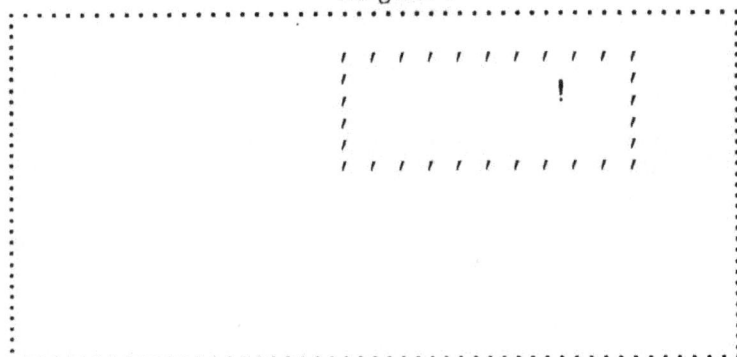

Humankind appears. The biosphere continues its invasion.

When we have the power to eliminate a form of life from the biosphere, a form that we do not know as wealth, we must face a truth; we have the power to eliminate a source of future wealth.

Humankind has no sympathy for smallpox, but smallpox is some of what the biosphere knows. We must ask, "What does smallpox know that we do not know?" From the point of view of the smallpox virus, its sphere of wealth (class-wsmallpox) shrank from a large portion of the biosphere, where it devastated human populations, to its imprisonment in a few isolation units. Our concern has shifted. Once we feared for our survival. Now we must fear loss of biospheric knowledge into which we might one day grow.

Diagram 5

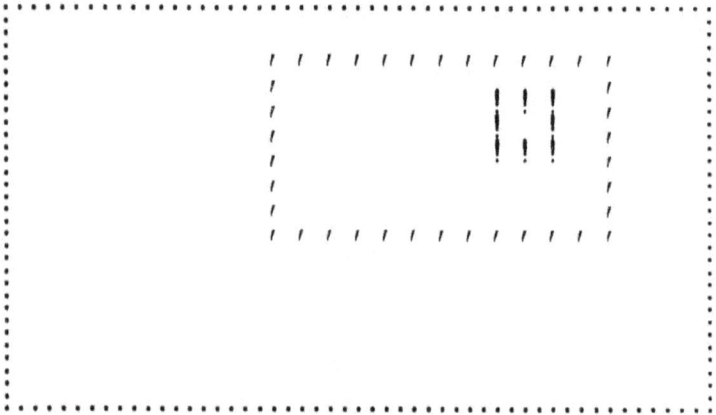

Humankind prospers. The biosphere continues its invasion.

Humankind is of the biosphere.
Biosphere is of the universe.

Life's challenge to the biosphere:
. become one with your universe.

Life's challenge to humankind:
. become one with your biosphere.
Meeting—or failing—our challenge
we will change:
. change in form
. change in method
. change in abundance.

‖

Diagram 6

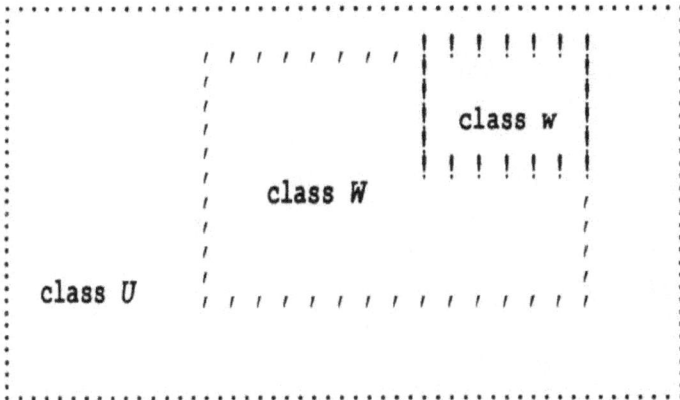

Humankind (**class w**) helps the biosphere (**class W**)
accelerate its invasion into the universe (**class U**).

13. Life's Imperfect Invasion of the Universe

Demon of today,
tormenting tester,
inhibitor of brave and bold,
a once hero
changed by subtle imperception,
you are that mistake we may make
or have just now made over
into more guilt and gossip.

. (The same mistake
. that we will someday remake
. into light hearted fame and folklore.)
Today,
you are the pains that we feel will last
that now make us cringe
and make us fear our self
as much too dangerously human.
In lonely race to uncommon disgrace,
we cannot face today's new imperfection.

Hero from our past,
delightful jester,
entertainer with tales of old,
a once demon
changed by time and new perception,
you are that mistake that time takes
and so slowly makes over
into more lore and legend.
. (The same mistake
. that we will each day remake
. with feelings of shame and sorrow.)
Today,
you are the aches from decades gone past
that now let us laugh
and let us accept our self
among strangely similar humans.
Together within our common old sins,
now we can grin at older imperfection.
About Our Acceptance Of Older Imperfection

In a lifeless universe, energy flowed into, around and out of many forms of matter. These multiple forms interacted with themselves, creating more complex forms. Amidst the trillions of trillions of interactions, energy combined in a pattern that could purposefully reproduce itself. This pattern could capture and organize more energy to sustain itself and repeat the same pattern. This was life. It may have started many times. Successful life needs a way to remember how to maintain and reproduce itself. In the form that we know, it is the genes in our cellular DNA that remember. All of life needs some type of genetic organization that tells one generation of interactions how to repeat the essential interactions of previous generations.

If a perfect form of perfect life ever lived
(before or since this life we know)
it fell to the perils of its own perfection.

‖

Perfect life reproduces itself infallibly. One generation is guaranteed to be exactly like the previous. Maybe this living perfection had no generations at all. One creature just continued. It grew; it ebbed and flowed; or it stayed exactly as it was. This perfect form of life had to dwell in an inconsistent world. Extreme changes happened in the dynamic universe and in each niche of the young and dynamic planet Earth. Unable to adapt, this perfect life form died and became inanimate residue of a perfectly beautiful but inflexible creature.

Long-lasting life must be imperfect. It works well most of the time, but its memory must fail often enough for some members of new generations to differ from their parents. Ongoing imperfection allows life to undergo adaptive evolution, so that, when the inanimate environment changes, some forms of life survive. Many die, but the chain continues. Life's imperfect invasion of universe continues.

14. Biosphere: A First-Class Wealth Organization

In setting forth his Critical Path for humankind R. Buckminster Fuller made a distinction between class-one and class-two evolution. Class-two evolution assumes humankind runs the universe. It includes the events that seem to us to come from human initiative and control. We like class-two evolution.

I planted a seed

here somewhere, where
today there should be
(by my rights)
a seedling-

here on time, when
the packet guaranteed
(to my sight)
a sprouting-

if not now, then
it had better be
(as I hope)
by morning-

here somewhere, where
it might have been seen
tomorrow
(by evening).

Seeds And Rights

Class-one evolution transcends our vision, and (as one finds out waiting for seeds to grow) it often voids our class-two-right to have things happen as we told ourselves they should. Class-one evolution came first and will leave last. It put humankind on Earth. Using R. Buckminster Fuller's words, Class-one evolution is typified by "ephemeralization," by which we do more and more with less and less, and by "acceleration," by which the time between changes gets less and less.

The biosphere and humankind continually organize and reorganize. Life grew in complexity and robustness by organizing itself. Simple forms of life organized themselves into cells. Some cells organized their reproduction so that instead of creating totally new and independent creatures they created specialized organs within a multiple cell creature. The invention of organs exemplifies ephemeralization, because cells working in a multi-cell and

40

multi-organ individual can often do more with available resources than can the same number of single-celled creatures.Since we need old fashioned cell-into-cell reproduction to make and sustain our organs, we multi-celled creatures need some enhancement to reproduce our entire organized selves. Class-one evolution gave some of us sex. Sex became an unwitting accelerator of evolution as we mix and re-mix the gene pool (inclined as we are to attract strange mates). New species and communities of species could now evolve faster. Evolution is subject to its own class-one evolution. It accelerates as the gene pool diversifies.

I wish to emphasize the obvious; the word, organization, comes from the root-word, organ. We say that people get organized (into committees and such) while their bodies have organs (heart, lung, foot, skin). It sounds fine to say that the human body is organized, and it sounds only a little strange to say that the police department is an organ of the city government. At the edges of evolution, then, organs and organizations form. Through class-one organization the biosphere develops, reproduces, modifies, and discards patterned interactions.

Class-one organization,
first-class organization,
is biospheric economic development.
Unwittingly,
we participate.
Wittingly,
but ignorantly,
our second class efforts
often retard or threaten our wealth,
our biosphere's wealth.
Our species is organ to the biosphere.
We are organs to our species.
Looking inward,
we see our organs
as their cells and genes organize them.

Looking outward,
we see ourselves
as parts in several organizations:
. family,
. neighborhood,
. team,
. school,
. country,
. ecosystem,
. bioregion,
. species,
. biosphere.
Organs work neither perfectly nor forever.
(Such is the trial-and-error of evolution.)
Organized, imperfect, timed interaction
among species,
between species and their environment;
call it ecosystem.
Around the planet
call it biosphere.
Across space,
someday call it bio-universe.

Biosphere sustains life and culture,
changes as inanimate environment changes,
changes as life alters itself.
Energy captured by biosphere,
organized into life,
organized to support life,
reorganized by life,
this energy is biospheric knowledge.
Biosphere knows
to preserve,
create,
re-create,
,to continue a habit of change.

Without such knowing,
life into lifeless ignorance,
will end.

‖

15. Capture, Distribute, and Recirculate Wealth

I divide wealth development into three processes that
interconnect in an evolving spiral. The three processes are:
capture, distribution, and recirculation. The following
diagrams from *Plum Local III* help me see how wealth of
biosphere and its parts can grow. Among the casualties of
simplicity, I ignore backward steps such as death and
disorganization. I will describe these diagrams as starting at
the origin of the biosphere, but, with simple changes in
vocabulary, they could start today and look at one person, a
tidal pond, a human community, an ant hill, a business, or a
watershed. As the diagrams progress, more energy of the
universe serves life, and life becomes more complex.

```
          Diagram 7
- - - - - - - - - - - - - ->
- - - - - - - - - - - - - ->
- - - - - - - - - - - - - ->
- - - - - - - - - - - - - ->
- - - - - - - - - - - - - ->
- - - - - - - - - - - - - ->
- - - - - - - - - - - - - ->
```

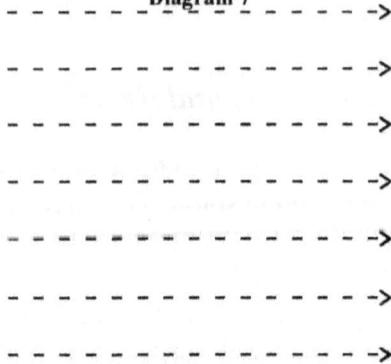

Inanimate energy. Energy distributes itself among dynamic but inanimate events in a universe that has no life.

```
          Diagram 8
- - - - - - - - - - - - - ->
- - G - - - - - - - - - - ->
- - - - - - - - - - - - - ->
- - - - - - - - - - - - - ->
- - - - - - - - - - - - - ->
- - - - - - - - - - - - - ->
- - - - - - - - - - - - - ->
```

A low-level organism. Among countless combinations and recombinations, some energy organizes itself into a simple living organism (**G**), which captures some of the passing inanimate energy, uses it, and then expels it back into the inanimate universe as waste.

Diagram 9

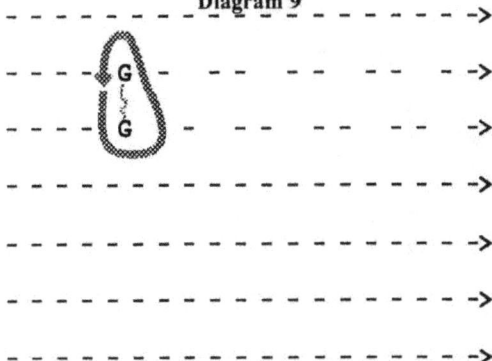

The biosphere, a high-level organism. Life survives
because it knows how to repeat itself —to reproduce. Two
creatures (organisms) now exist where there was only one
—at least doubling life's capture of universal energy. Almost
immediately, these organisms interdepend —cooperating and
competing in a patterned way for the same space and food.
Each creature accidentally recirculates some of its energy to
benefit the other. This interdependence between low-level
creatures is itself a high-level organism:
The biosphere is born.

Diagram 10

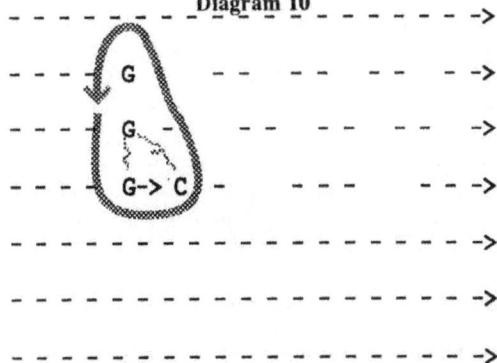

More low-level organisms. Again, the growing biosphere
captures more energy. Repetition occurs twice, once
perfectly, once imperfectly —creating a somewhat different
creature (C) that finds a niche consuming waste energy
distributed to it by (G). Thus, the energy captured from the
inanimate universe stays longer in the expanding biosphere.

45

Diagram 11

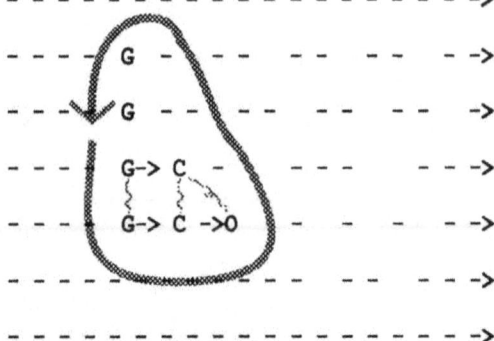

Again more low-level organisms. Repetition continues, twice perfectly and once imperfectly. The biosphere captures yet more energy. The newest low-level creature (0) can use the waste distributed to it by the older ones. The biosphere again increases in depth as some energy stays even longer before returning to the inanimate universe.

Diagram 12

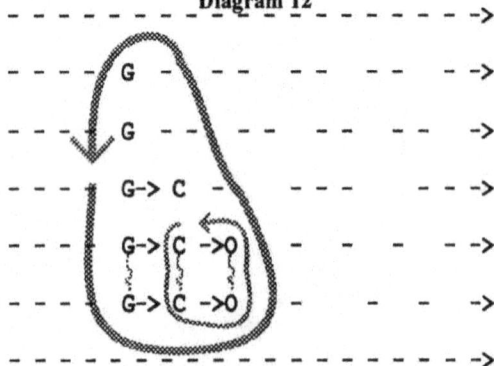

A mid-level organism. Repetition continues —perfect three out of three times. The biosphere again captures more energy. More gets distributed in depth. More gets recirculated back up the chain of distribution. A new interdependency has developed among some (C)s and (0)s, creating a mid-level organization: ⌒⟶ The new level of organization could be a tribe, a ⌣ family, an ecosystem, a kindergarten; or a separate, multi-organ creature such as an ant, a person, a tree.

The number of creatures increased tenfold between diagrams 8 and 12. The amount of energy in the biosphere increased more than tenfold because energy captured from the universe gets distributed and recirculated before leaving the biosphere at a rate somewhat slower than it enters.

16. Biology, Brain, Artifact: Our Wealth Stores

Where do we store our wealth of knowledge?
Evolution gave us a nervous system,
a brain, wherein we learn our environment
and attempt to respond to change.
Our brain holds memory
wherein we store our learning for later use.
We learned to create artifacts,
utilitarian and coded,
that, once made,
hold our knowledge.
Nests and tools,
paintings and hearths,
books and buildings
may survive our death
to be inherited by following generations
that may use and build upon this heritage.
Our life-time learning,
our artifacts,
and our biological heritage
combine to organize our culture.

‖

We can classify human knowledge by where we store it. Carl Sagan wrote of genetic and extragenetic locations.

Genetic knowledge is written in the nucleotides in the chromosomal DNA molecules. As life evolved its forms became ever more complex. Genetic code had to contain more information. Humankind has more genetic information than most other mammals, which have more than amphibians, which have more than protozoa. Depending on mutations and natural selection, this genetic form of knowledge grows slowly and is limited by its container, the chromosome.

Extragenetic knowledge resides outside of our chromosomes. Much is written during our lifetime in the memory of our brain. Because we can learn and be taught, our species can adapt quickly to problems and opportunities presented by our environment. But brain knowledge dies with each individual unless she has somehow passed it on to someone else. It is still sharply limited by the capacity of a brain to learn and apply information during one lifetime.

Some extragenetic knowledge, however, is also extrasomatic; it is stored in artifacts outside our bodies. Utilitarian objects and coded messages augment our ability to capture and use energy. Humankind does not have exclusive access to such knowledge, but we are the experts. Libraries are obvious examples, but tools, houses, and other useful objects contain knowledge. These inanimate objects may have a usefulness that is independent of the life span of the people who create them. The objects tend to accumulate, offering each successive generation an opportunity to be more knowledgeable (wealthier) than the previous one. Nor is it necessary for the user to know as much as the creator.

To use our hammer of steel and wood
we must know to grow and manage arm and hand.
We must learn how and why to wield our hammer.
Yet, we need not know how
to make or shape steel,
to select wood
nor make a handle.
Someone put this into our hammer,
so we can use without knowing.

Someone Put This Into Our Hammer

I classify knowledge by location as: biological knowledge, brain knowledge, and artifactual knowledge.

Biological Knowledge. It includes the genetic codes, most of the living tissue of each living creature, and any energy immediately at the disposal of this knowledge. For an earthworm it includes himself, the soil that surrounds him, that is about to be consumed, and that that is being consumed. For the biosphere, it includes the biomass plus most of the inanimate materials and gases just above and below the Earth's surface, and beneficial energy arriving at Earth from the sun and elsewhere.

Brain Knowledge. It includes information and skills learned by individuals from their environment during their lifetime as well as any energy immediately at the disposal of this knowledge. For the human who has learned to swim and dive for clams, it includes the water supporting her and the clam she is about to grasp. For the biosphere, it includes any increase in biomass or captured energy that is due to brain knowledge -for example, the fish stocked and propagating in a natural lake that had no fish until knowledgeable humans intervened.

Artifactual Knowledge. It includes knowledge outside our bodies, and any energy immediately at the disposal of this knowledge. For a cold human it includes his stove that incorporates brain knowledge of metallurgy and the science of burning. It includes the firewood in and immediately nearby his stove, as well as the food he is about to cook, and his cookbook. For the biosphere, it includes a man-made fish pond, the solar panels on a space vehicle, and the energy that each captures.

Knowledge organizes and reorganizes the world. Biological knowledge sustains basic processes and carries the fundamental codes of life. Brain knowledge augments the amount of energy that an individual, a group, an ecosystem, a species, or biosphere can put to beneficial use. Artifactual knowledge can harness and release vast quantities of energy that the body and mind cannot manage directly.

Brain knowledge increases rapidly for each new individual. An ignorant baby becomes a knowledgeable adult. We organize ourselves so that we can pass much wealth from individual to individual and from generation to generation through example, oral tradition, ceremony, song, dance, and vocabulary. Much brain knowledge also comes from direct personal experience. This wealth accumulates much faster than does genetic change, but it is inefficient. It usually passes to only a small group before the teacher dies. It may be forgotten or misunderstood. Much brain wealth dies with each individual.

Artifactual knowledge increased slowly in the early millennia of mankind. One generation left simple hand tools in wood, stone or bronze for the next generation. Supplemented by a transfer of brain knowledge through teaching and demonstration, these artifacts often retained their usefulness over generations. Each new generation could not only make new tools but use those accumulated by earlier generations. When nomadic groups converted to stationary communities they could keep more of the tools made by past generations. We still use simple hand tools. They often help us build complex machinery and electronic equipment that in turn we can use to manufacture better, less-costly simple hand tools. Our artifactual culture builds upon itself.

Written language is itself an artifact (in code) that helps us store instructions for using and building other artifacts. Instructions might otherwise be forgotten or severely limited in distribution. Using this coded artifact, our brain knowledge can quickly and temporarily grow to suit the need at hand. We only need to know how to read the language.

I illustrate the interworking of the forms of knowledge with a simplified view of the ancient Egyptians and their culture that flourished along the Nile River. They relied without thinking about it upon their biological knowledge to keep their bodies functioning long enough to reproduce themselves and to accumulate and transfer brain knowledge.

With their brain knowledge they could understand the seasons, the floods, the principles of agriculture and mechanics. Combining biological and brain knowledge, specialized engineers could build dikes and irrigation systems. These artifacts enabled many who knew nothing of engineering to increase their food and fiber production. In the process, Egyptians developed new strains of plants by selecting seeds from those plants that took best advantage of this man-altered growing system. These plants, though having their own genetic wealth, were living artifacts storing extrasomatic knowledge for the ancient Egyptians and we who follow. A complex system of biological, brain, and artifactual knowledge built an increasingly productive culture that gathered ever more energy unto itself.

In a purist sense, the three forms of knowledge are downwardly dependent. Biological knowledge came first; it can exist without the more advanced forms of knowledge. Brain knowledge evolved with increasingly sophisticated nervous systems. It obviously depends upon the genetic code of the body that hosts it. Artifactual knowledge exists outside the body. However, without brain knowledge and biological knowledge there would be no one to take advantage of it. It would cease to be wealth.

This hierarchy of dependence is valid, but difficult to apply. Much biological wealth depends on brain knowledge or artifactual knowledge. Pigeons can exist without humankind, but their population would be much lower without the artifacts of man, city buildings. Holstein cattle owe their unique characteristics to careful breeding by humans. They are living artifacts for humankind, but biological wealth for themselves. Astronomy is mostly brain knowledge, but much of it depends upon artifacts such as telescopes. In our complex patterned biosphere the forms of knowledge interact and interdepend.

17. Culture: Patterned to Cooperate and Compete

Energy begat matter.
Matter begat life.
Life begat knowledge.
Knowledge begat culture.
Then culture begat.

Begat

Life is a nearly inevitable pattern that emerges from changing interactions among inanimate forms of energy. Adaptive evolution is a nearly inevitable pattern that emerges from life interacting with itself and a changing inanimate environment under an imperfect system of reproduction. Culture is a nearly inevitable layer of patterns that emerges from adaptive evolution interacting with itself. Complex patterns of life emerge and get tested by adversity and diversity. The more durable patterns survive and retain knowledge as to how groups of species and groups of individuals can organize to exploit the inanimate and animate forms of energy in the environment. This group knowledge is still stored inside individual chromosomal genes, but it may be used in patterns across the genes of several individuals and even multiple species. It may be used in patterns alive in places that range from virus to atmosphere.

Richard Dawkins calls the patterned and mutually beneficial relationship among individuals of one or more species, the "Extended Phenotype" I think of it as the larger body. Congress is a lawmaking body. The actions of its members are hard enough to explain when we know the role of Congress. Certainly, we would have no idea what single acts of individual congressmen mean if we did not see the larger body.

Dawkins argues that the selfish genes that drive evolution do not just affect the creatures in which they reside. Behavior will tend to maximize survival of genes that foster the behavior, whether the genes are in the animal behaving or in some other creature that affects its behavior. Whenever cooperation or competition increases the survivability of genes (in separate species, separate individuals in the same species, or separate organs within the individual) such behavior is reinforced.

Evolution creates not only diverse species that fit diverse habitats, but also creates diverse extended phenotypes, relationships and cultures that fit diverse circumstances. Humankind competes with the cold virus, yet Dawkins asks whether it is we or the virus who has manipulated the evolution of our sneeze in response to a cold. The sneeze provides us some relief, but clearly the virus is given a free ride toward other victims. We and the cold virus have a special relationship.

Thus, the distinction between cooperation and competition becomes fuzzy. While individuals act out serious competition to improve their well-being, they may play compatible roles in a cooperative effort, a cultural effort, to survive. Economics should be one of the sciences that try to understand culture—to understand how we use patterns of cooperation and competition to improve well-being for individuals and increase likelihood of survival for our species.

18. Exploitation's Deceitful Attraction

We know from experience that, despite all the good it does for us, at times Adam Smith's "invisible hand" needs to be slapped. We rise again and again in fits of moral outrage to knock down those who sacrifice common good to disparity —promote our loss for their gain. Do we do so often enough? We discuss that question frequently. Here is another question that we should discuss: do such struggles and knock downs pit morality against economics? To this second question I answer, no. Morality need not fight economics unless you equate economics with pursuit of personal gain. You should not. I do see these struggles as contests between good and bad; but between (good) morality and bad economics, between common good and bad science. Great disparity is a great clot in the arteries of development.

Great disparity between us,
great impediment among us.
Moral justice asserts
that our wealthy must distribute wealth
onward to our poor.
Economic development just begs
that wealthy just act
knowing their dependence upon the poor
—and upon more.
More wealth distributed
brings back more wealth recirculated;
today distributed,
tomorrow shared.
In abstract we speak
of wealthy and poor.
In concrete we speak
of butcher, baker, banker,
slave, master,
pitcher, raker, candlestick maker.
We distribute,
recirculate,
compete,
cooperate,
evolve,
to survive
to prosper
together.

||

Consider the economics of exploitation. Take an extreme example; consider the economics of slavery. As species, as group, our wealth grows biologically with each birth; it decreases with each death—usually. Simply having more people may decrease our group wealth. Our increased population may overwhelm brain knowledge and artifactual knowledge that we use to harness resources. While this negative outcome may not be as likely as doomsayers often say, more people do not by their presence alone make the individuals in their group more wealthy. Yet, mistaken actions by fellow humans seem to stem from a belief that greater population brings wealth—if it is more of the right kind of population, the kind that makes us rich. When powerful individuals and groups enslave other people they increase the population that works at their service. A slaveholder feels wealthier holding more slaves.

We can exploit others in ways more subtle than slavery, but in all ways exploiters believe (if they are honest with themselves) that they become wealthier if they subjugate more people. Usually they do not try to increase overall population, just their controlled population. Unfortunately, history shows that from their narrow and short-sighted viewpoints they are usually right. Since we see our world having static resources, we find it selfishly attractive to exploit the minds and bodies of our brothers and sisters. These harnessed minds and bodies know how to gather wealth for us.

From the broad viewpoint of humankind exploitation is not attractive. It reduces opportunities for the exploited. That reduction inhibits the growth of brain knowledge and artifactual knowledge for our species, reducing species wealth and average per-person wealth. To the exploited it is painfully obvious that they suffer the burden of their exploiters. Unfortunately, exploiters are painfully ignorant that they suffer under the burden of their exploitation. This becomes tragically obvious when a lethal microbe flourishes in the impoverished ecology of the exploited and surges forth to kill exploited and exploiter alike. Yet the lesson is not learned. Here, economics has, first, much to learn and, second, much to teach.

Exploiter dreads that exploited would self-organize.
Old culture that slave might keep,
and new culture that slave might create,
destroyed, discouraged, undone.

Exploiter simplifies—
in
simple
organization
finds
control.
Biosphere complicates—
for in our complexity, it finds resilience;
for in its resilience, we find wealth.

||

19. We are Cursed and Blessed.

Life reproduces itself, but imperfectly. This imperfection creates an intricate, diverse biosphere.

Humankind has two traits that distinguish us from other species. I believe these traits contribute to our unique role in the biosphere.
We have a brain that has somewhat independent hemispheres in which we process information in at least two different ways at the same time. We have social systems that tend toward stability, but in which we often and inevitably undergo instability.
We seem cursed by the anguish of war between the conscious and subconscious and between the logical and intuitive factions in our brain. We seem cursed by bloody and bloodless wars between factions in our society.
Other species evolved with greater social order.
Compared to our own, their brain parts seem to work in harmony. Our unstable society may well be a curse brought on by our unstable brain. Yet, our creative ability to explore the workings of the universe through art, science, engineering and fantasy is a blessing that we also owe to that curse. If there is to be a utopia, we must find it in how we manage our continuing instability.
All species can perceive scarcity in resources, and they see an apparent need to manage scarcity by preventing

any change that instability will bring. We, alone, have the blind management power, at this tick in the evolutionary clock, to destroy life and its biosphere of change.
It is also only we who can hope to notice, midst war and destruction, the wealth that instability and change have wrought within our biosphere. Only we can hope to see that beneath the apparent rules of scarcity and competition lie more general rules for us to follow-rules for abundance, and rules for cooperation.

Blessed Curses

Humankind seems cursed by its heritage of imperfection in ways that other species are not. Yet, we are blessed with a preeminent role in the biosphere. Richard Restak described well how our semi-independent brain hemispheres process two ways of thinking that are generally coordinated but often conflicting. This duality is mirrored in our culture. We experiment with: democracy, dictatorship, republic; polygamy, polyandry, monogamy; matriarchy, patriarchy; monarchy, anarchy. Only the human species has a restless urge to understand and perfect.

Competition seeks instability.
One individual,
one group
tries to gain advantage over another.
When too successful,
competition fails.
It can produce a stable dictatorship,
monopoly,
slavery.
Cooperation seeks stability.
It too can fail
if it is perfectly successful.
Stable wealth stops evolving.
Rebels appear.
Feeling stifled,
they promote instability.
We play hopscotch
through patterns of opposites:

cooperation or competition,
with
stability or instability.
War -unstable competition.
The marketplace -stable competition.
Tradition and rules -stable cooperation.
Invention and information,
though often spawned by cooperation,
bring change -instability.
Each pattern brings the other.
The market creates incentive for invention:
a better mouse trap,
a five-cent cigar.
Invention produces new traditions:
stay home
to watch a ball game at the park,
go to the park
to get away from my computer-cottage office.
A stifling tradition brings revolution and war.
We are special
because we can notice these patterns,
seek to understand them,
and decide to participate.

What Makes Us Special

If an objective visitor from outer space had arrived on Earth a few hundred thousand years ago she would have had to stay around to notice that we naked apes were in any way special. Today, she would quickly pick us out of the crowd. We have made ourselves quite noticeable. Electric hair dryers, tennis shoes, oil spills, and political conventions make us stand out clearly in the crowd of earthly creatures.

We see correctly that we are special, but we have tended to think incorrectly that we are special because we are better. "Better" is a subjective evaluation, but if we equate it with "more perfect" we are wrong. We are special because we are less perfect. We are cursed with imperfections that bless us with flourishing culture. Individually and in groups we play dynamic roles that never allow our species to settle into a comfortable niche.

20. Family is Our Minimum.

There are no true rugged individuals.
Either they died
for lack of nurturers after birth,
or they were never born
for lack of lovers before conception.

Ruggedless

The science of ecology tells us how we share biospheric wealth with other species, often to our mutual benefit. We must share. No element of life is self-sufficient. We must tax other individuals and species; they must tax us. Humankind has no self-sufficient individuals. From Lewis Thomas in *Lives Of A Cell*, I learned that the mitochondria in our cells are but one of several kinds of creatures with their own genetic structures that live symbiotically with us, within us. We tax them for their knowledge in order that we might survive; human genes alone do not know enough to keep us alive.

Even if I were to refuse to recognize mitochondria as anything other than "me," I could not ignore the two people who had to undergo bisexual reproduction for me to arrive in this world. Then I cannot ignore that once I got here, during infancy and well beyond, I continued to depend upon adult nurturers for my survival and well-being.

To support bisexual reproduction and to solve the challenge of infant dependency we organize ourselves into families. Family is the minimum biological unit of organization for species survival. Many of us do more than survive. We depend upon our families and their particular patterns of role playing to do more than reproduce and nurture hunks of genes and protoplasm. Adults carry brain knowledge and have artifacts that they wish to distribute to their children and others. Family is the convenient minimum to carry out this complex wealth distribution. Through family we receive and pass along culture and experience. Thus, our adopted children—who carry none of our genes but carry knowledge that we acquired during life—may inherit and pass along more of what we are than do our biological children.

Family is the minimum—the macroeconomic minimum—the smallest organ that our species knows can capture, distribute, and recirculate knowledge we need to survive and prosper. Since reproduction requires two people and produces a dependent, over time there must be at least three individuals in a family. The minimum number of roles to play, however, is four: father, mother, dependent, and nurturer. There are no maxima.

Beyond the biological minimum necessary for survival, our species has unique flexibility in choosing the family form through which one generation can pass biological, brain and artifactual knowledge on to the next. Adjectives such as matrilineal, patrilineal, extended, monogamous, and polygamous each describe a family pattern that has proved useful. Nouns such as nucleus, band, clan, tribe, chiefdom, city, federation, and even nation, describe scales at which family-type roles can operate.

Microeconomics is important because it helps us understand the dynamics of individuals (or even families acting as individual entities) who buy and sell goods and services in a marketplace. Some of these individuals may be close to rugged, but, here at least, we do not give a damn. What does it profit a man to control a market but suffer a dearth of things to trade? We cannot look to microeconomics to understand the dynamics that produced the market goods and services in the first place -nor the greater wealth that never goes to market. To gain that understanding we should be able to turn to macroeconomics and its minimum economic unit, the family, in all its forms. There we can hope to find out how our species reproduces, expands and nurtures its wealth.

21. *From Family to Government*

Parents in our nuclear family, our elders in our clan, our council in our city, our nation's government—each makes decisions for the benefit of one of our groups. National government differs in scale and organizational complexity from a family council, but not in function. Each represents an organized effort to promote role playing by group members. Members, citizens, hope that roles are played well enough to improve group and individual well-being. We are members of several groups; several decision-making bodies represent us. As we invent and join new groups many old ones persist. We happen to call a class of broad-reaching groups, or their representatives, government.

We have such propensity to form government in our groups that among our deeply embedded cultural patterns there must be one that causes us to produce and reproduce government. Provided we remember that all genetic action must occur within our true cellular genes, we can think of cultural genes for government.

These genes are persistent yet imperfect. Some stable societies evolved with stable forms for government. Yet, as our species flourished and spread its influence around the globe, new circumstances tested old forms of organization. Changes had higher likelihood of survival if they enabled the group to better invest group knowledge. Government gains respect among constituents and also among rival groups if it enables the group to organize itself to produce and protect more knowledge than it invests. Such successful government is likely to be imitated—just as successful biological parents are imitated.

Pick any important company in a marketplace country such as the USA. Do a thought problem. Consider the impacts on the nation if we eliminate that company. Pick any whole sector of a market economy and do the same thought problem. Choose the farming sector, for example. Eliminate it—a disastrous thought.

Now do the same thought problem, but eliminate only
government. If that is too difficult to imagine, eliminate
only governmental protection of property, and
governmental guarantees of public rights-of-way.

Government is family,
the industry we cannot do without.

||

22. Communal Commitment, Private Flourish

Government organizes group knowledge in ways that
individuals cannot or wish not.

Government can create a system of defense that is more
effective and occupies fewer people than if everyone were to
try to defend himself. Without this system the group might
be vulnerable to attack. Productivity could suffer because too
much time is spent on defense and fear.

Government can guarantee food and shelter to someone who
wishes to experiment with a new system for food production
—a system that could benefit the whole group. Government
induces a temporary flow of wealth away from the rest of the
group to the experimenter. If the experiment succeeds, that
wealth that had flowed away from most of the individuals in
the group will return to them with dividends.

If in our family
we lack communal decision and commitment,
private life does not flourish.
If in our government,
we lack communal decision and commitment
private enterprise does not flourish.

||

23. We Tax Ourselves When We Play Our Roles.

We cooperate; we compete. In both modes we exhibit freedom. At least, that is a view we can choose. In another view, we just play our inherited roles. "Role playing," "role differentiation," these are terms frequently heard among social scientists. They are also better and more general terms for what we and our economists call "taxation."

Role playing (1) distributes wealth among individuals, and (2) creates wealth when the roles are mutually beneficial. Number one is well understood by both charitable and exploitive people, by our tax collectors, and by other species —the more I get from you, the better off I am. Attribute number two is at the frontier of our understanding of macroeconomics.

During the dawn of life separate organisms that were accidentally different accidentally contributed to each other's chances for survival. Some produced offspring that had a tendency to organize themselves to reënact the ritual. As they did, role playing became part of the biosphere; taxation became part of the biosphere.

Each particle of DNA, each species, each individual, each group depends upon and is depended upon by other elements of the biosphere. Each taxes others and is taxed in turn within an intricately organized web of roles. On occasion, the quantity of life traumatically decreased. Despite such setbacks, life incorporated an increasing portion of universal energy into itself and into its complex support systems. As life grew, so did the complexity of the roles played by the organs and organisms that sustained it. Had we been there, at some point the complexity would have become too much for our great brains to comprehend. We would have to develop concepts of freedom, intention, randomness, in our attempt to understand cause-and-effect in what we saw. We would find words for cooperation and competition. In some order (perhaps simultaneously) science, pseudoscience, religion, and philosophy would follow.

As life flourished, new roles emerged, old roles changed. Through its organized role playing the biosphere learned not only how to prevent a net loss of biological energy (absent great trauma), but also how to capture more energy. In terms of biosphere economics, wealth steadily increased. While some species lost all wealth and became extinct, humankind prospered. Human economic history, when read from a distance, records changing patterns of roles that were adequate to keep us around and increase our wealth.

Role playing works because it distributes resources among individuals to the benefit of the group. An infant needs nurture, so it taxes its family to supply energy and skills to survive. Given survival, the infant can go on to contribute to the welfare of the family and to help produce another generation. Our tendency to nurture others lets us help individuals become scholars, artists, scientists and administrators. It causes us to make an effort to preserve life and opportunities for less fortunate people even when we seem to loose wealth in the process. We harvest and create food, education and security. Then we tax ourselves. We cooperate and compete so that we distribute these products and services among us. History is too full of examples of exploitation and persecution. We live with the guilt that comes with our unfortunately large margin for error. Nevertheless, so far, we tend to tax ourselves well enough to survive and occasionally flourish.

24. Property is Acknowledgement.

I have a complaint and a confession. Complaint: when we talk about "private property rights," we show that we do not know what we talk about. Confession: I am guilty of having used this puffed up phrase. I hope I have cured myself. Let me try to cure you.

Why does the phrase, "private property rights" show ignorance? That term is twice redundant—long-winded. Property is not land, it is not a book. Property is a right to be private, to priváte, to deprive others of access and use. So, to speak of "private property," of "property rights," or of "private property rights" is to speak only of property. If we can argue without this redundancy perhaps we will be short-winded. Perhaps we will better understand property.

Ah, you may say, but what about "public property," the opposite of private property? It is no opposite, but rather an oxymoron. If you insist on using the term, its only meaning can be "non-property"—nobody has the right to restrict our access and use. A sunset is beautiful public property; better, it is a "public good" in the "public domain." Your county land fill is not public property; it is county property. The county has rights over access and use of the land fill as would an individual owner.

Are these rights absolute? Whether held by you or the county, property is within a higher domain, a pre-eminent domain, an eminent domain. "Eminent domain," as we use the term in U.S. America, includes the obligation by government to pay justly for property, but it is also a self-evident declaration that property is a guest within the higher domain of society. Complaint over!

We can discuss individual wealth because individuals can control and trade goods and services. However, this power to control and trade is always delegated by a larger group. Property is group acknowledgment that an individual can restrict access to something—a by-group-acknowledged right to be private, to priváte, to deprive others of access and use.

Acknowledgment may be given grudgingly to someone who has taken by force; property is not necessarily untainted. If the group refuses to acknowledge ownership, and the possessor must continually defend her possessions against others, she has booty—not property. She cannot freely share, trade, sell, give, or lend her booty. If the group were to choose to acknowledge her possessions as property she could take them to the marketplace or leave them at home and expect community assistance if someone tries to steal them.

The marketplace is a group sanctioned center of activity. We can say that people go there to trade property. In a marketplace, operating under rules established by the group, people trade group-acknowledged rights of restricted access. An individual restricts access to her possessions until someone else gives her the right to restrict access to items that she believes to be at least of equal value. When those rights are exchanged, property is exchanged. The exchange is complete when society acknowledges that property has changed hands.

25. Back In 1978: a Myth

Exclusive Report on the Minnehaha Conclave

As the 1980 elections approach I can't believe that I am the first to break the seal of secrecy over the events of Halloween, 1978—the last days before the last national election. I hope that by revealing what happened that night I can save others the tragic disillusionment from which I am now almost recovered.

There were so many conservative Republican candidates,conservative Democratic candidates,and "don't-tread-on-me" independent candidates trying to get elected and dismantle the government that I decided to seize the opportunity.

I awoke in a sweat. I was trying to get over a nightmare about having a secure government job to turn to if all else failed. It was then I decided to put out a call to a secret conclave.

I jotted down some brief but detailed instructions, telegraphed them to the politically astute friends we have left in the wake of our peregrinations across the country, and took off to make preparations at the conclave site.

Taking a hint from the College of Cardinals, I ordered up a ton of Ritz crackers and a thousand bottles of mineral water. These would serve as our only sustenance until the conclave had done its thing.

Then it began. While kids across the country were home getting sick over candy they had been collecting all evening, while voters who had taken one night off from eagerly attending political rallies were relishing the tricks they had played on the little pests in their neighborhood, while all this was happening, the disguised candidates began arriving in procession to my conclave—each with one lighted candle in hand and an attaché case filled with extras.

Anti-government candidates are to-a-man and to-a-woman punctual. Thus it was, that by 11:33 p.m. central standard time, in the freezing cold confines of the narrow canyon formed by Minnehaha Falls in Minneapolis, 315 conclavers were assembled. Like choir boys in down parkas and Lone-Ranger-type Halloween masks, they were ready to go to work.

I took 27 minutes for my opening remarks. In sum, I told them of the inevitable victory of their cause. It would soon be within the power of this esteemed group to create a utopian country that had neither government nor bureaucrats. The people of the country were speaking. These 315 men and women were the best listeners.

Only two necessary elements for the ultimate victory were missing.

One missing element was their own combined conviction that the holy mission on which each had individually embarked could be accomplished if the conclavers were willing to march as one in a crusade. "Are you ready for the crusade?" I whispered emphatically. My question was answered by four minutes of delirious shouting, embracing, hand shaking, and the chant: Hit'em again, hit'em again, harder! harder!

After joyfully raising my arms to restore calm, I told them that the second element missing was a simple program for implementation. It could be a simple program, because all that is necessary is to ease the bureaucrats out of their offices and let private enterprise bid on the vacant facilities. A few other details in the program needed to be worked out, but it could all be planned by sunrise -in time for everyone to get back to the campaign.

After all, the program wasn't so terribly necessary in itself. We would just have to be ready with good information and public relations to head off the self-serving attacks that would be organized by the old guard and the socialists.

It was now midnight. The conclave set about its task.

Each of twelve sub-conclaves, whose members were chosen by lot, was to produce one paragraph for our program and one for our proclamation.

I passed out Ritz crackers and fielded general questions from the various sub-conclaves.

...

"Hey, our group decided to keep our armed forces and the police. Otherwise, anarchists and communists would run roughshod over us. Is that all right?"

"Sure, but figure out how to administer them without bureaucrats."

"We're not sure whether Greyhound, Brinks, or General Motors is capable of taking over the street and highway system. Does anyone have any ideas? ... O.K. We'll keep working on it."

"Some of us westerners were wondering if it would be possible to rewrite some of the history books our kids are forced to read in school. That bit about the government dividing up land, giving it to our grandfathers, and building water projects, and subsidizing our production doesn't sound too good for our cause."

"Don't worry. Group 7 is doing away with the schools. You'll be able to keep your kids at home and teach them what you want."

"We're a little worried about the possibilities of revolution if we eliminate all of the health, welfare and civil rights programs at once. Do you think it's all right to increase the army a little to keep everyone in check, or to phase programs out slowly so no demagogue can come along and incite riots?"

"Hmmm, well I'm afraid that if we did the latter, our own supporters would call us the Wishy Washy Conclave, rather than the Minnehaha Conclave."

"Hey, if anyone objects to having a cell in the basement of each home to handle convicted criminals on a rotating basis let us know. That's the way we are thinking."

The Ritz crackers were consumed by two o'clock. The questions ceased by 2:30. The rumble-rumble of their voices put me to sleep.

I awoke with the first rays of sunrise to find the whole conclave standing over me. They looked tired. They also looked a little discouraged. Yet, many seemed ready to burst into a smile. Then, one of my southern conclavers drawled, "He'ah it is."

I was puzzled. The piece of paper he gave me in the dim light was clearly a page torn from a book. He and each of the other conclavers patted me on the back as they filed up and out of the little canyon. Each said something to me like: "We did it," or, "That's just the first dozen lines, but you can fill in the rest."

I began to get t he idea. When I glanced down at the page I was given my suspicions were confirmed. The light was now just bright enough for me to read the small printed words:

> *"We the people of the United States, in order to form a more perfect union, establish justice ..."*

The next thing I remember is my wife reviving me in a pleasant cell at a Minneapolis police station.They had called her in from Colorado. This Halloween goblin, who appeared to be her husband, had been found skipping merrily through Minnehaha Park singing,

Kinky Commie Copouts!

Kinky Commie Copouts!

Kinky Commie Copouts! ...

Back In 1978

26. Capital and Social Isms

Civilizations have disappeared almost overnight when a governmental structure collapsed into the hands of people who could not or would not hold and propagate the communal knowledge nurtured by the government. The overthrow may have been justified by the abuses inflicted by those in power. Unfortunately, knowledge stored (possibly hoarded) by the abusers often got lost to future generations.

When a species or ecosystem disappears, an organ of the biosphere dies; wealth disappears. The biosphere no longer knows how to do some things that it had known. Wealth also disappears when a flourishing society dies.

Societies that we choose to call "primitive" survived with little change in technology or in their systems of government. Across the breadth of the species, however, somewhat unstable groups, "advanced" societies, experiment with ways to organize family, to invest group wealth, and to govern. Using our blessed curses, imperfect mind and imperfect reproduction, they extend evolution out of our bodies into society, to our family and government, to our social-isms and our capital-isms.

Ism-Capitál
     Ism-Commún
          Ism-Sociál

Planned Economy
     Marketplace
     Private Enterprise
     Free Enterprise
     Government Enterprise

How can I speak or write
these terms that have
          economic meanings
          political overtones
          historic undercurrents
          redundant redundancies
          stark contrasts
          subtle distinctions, and
          apparent similarities?

I want two of them to stand for all,
          to enter the ring, and then
                    to do battle.

At the edge of the ring paces Marketplace with
free enterprise in her soul
capitalism in her limbs, and
private enterprise in her brain.

In the center of the ring
revolves Socialism with
          communism for his soul,
          government enterprise
          for his body, and
          planned economy in his mind.

They spar but seem stuck in their zones.

71

They jab,
but their gloves barely clash
in the large space between.
(or do they exchange something
that I cannot see?)

She lunges to the center.
He feints and dances to the edge,
barely brushing
as they pass each other by.

Now, she occupies the center
while he patrols the edge.
But, somehow,
in making their moves
they have quick-changed costumes.

She now revolves
in the cloak of socialism
He, an equal transvestite,
paces in the dress of marketplace.
I don't need X-ray vision to see
that they have also interchanged their
>   *bodies*
>   *minds*
>   *limbs, and*
>   *souls.*

Apparently, some such parts
work well only at the edge
while others work well only at the center.

I had hoped to see a fight in the ring.

I moved ever higher and farther away in the arena
only to see it filled with rings of many sizes and shapes—
the edge of each one sharing the ropes of several others.

Each had its chief of the center
and one or more others
holding forth along the ropes,
where activity ebbed and flowed,
sometimes frantically.

What folly,
to think I could put both transvestites
into the same clothing.
Tragic were the fates of many rings
that did remove marketplace
from their edges,
as peaceful sport erupted into
thievery or war,
or disrupted into
wall or crevasse.

Extinct are the rings
that poisoned their communal centers—
their centerless edges
(having lost all definition)
rewoven with indifference into
the
ropes
of
other
rings.

Equal Transvestites

I

find it difficult to write about our "isms": capitalism and
socialism. Political and religious groups of divergent
persuasions have usurped and shaded the key vocabulary. A
simple sentence gets translated by the reader into his own
language, which may or may not coincide with my own.
True to some degree of all communication, I think the
reader-as-writer difficulty is a serious problem when
exploring marketplace and socialism, or capitalism and
communism. I believe that part of the reason why these

words have become burdened with emotion-filled connotation is that the economic definitions are themselves inaccurate.

I choose to discuss marketplace and socialism. "Marketplace," as I use it, includes the general concepts of capitalism, free enterprise, private enterprise, and other near synonyms. Of the choices, "marketplace" says the most and carries the least emotional baggage. For the other half of the discussion I choose "socialism." I use socialism to include communism, planned economy, government enterprise, and several other terms that express communal activity. However, the choice is not easy, so I use a mixture. "Community" and "society" are equally good words. I prefer the "common" and "communal" from "communism" to the "social of socialism," but I prefer the economic meaning of "socialize" over "communize."

27. Confusion over Capital and Ownership

During the Cold War which pitted U. S. America and its capitalist allies against the Soviet Union and its communist allies, capitalists confused economic communism with totalitarianism, egalitarianism, and atheism. Communists confused economic capitalism with colonialism, elitism, and self-righteous exploitation. Each group confused itself with values such as freedom, hope, and progress. Both relied, in part, on the classic definition that makes socialism the opposite of capitalism.

According to this classic definition, under socialism society should own the means of production (that is capital); under capitalism, private individuals and companies should own the means of production. Means of production usually include resources such as fuel, minerals, and lumber, as well as key industries that process and transport these resources. The definitions seem simple and intuitive, but only if we ignore some obvious questions.

How about eyesight and the opposable thumb? Where would most human production be without them?

How about the mind, the body, libraries, air, organization, and rights-of-way across space we do not own? We cannot produce without them.

We cannot produce without many forms of knowledge found in biology, brain, and artifact. Necessary resources get ignored in lists of means of production. Society cannot own mind or eyesight. Yet, to practice socialism under the classic definition, it must. For society to be truly capitalistic, private enterprise must own my body and our air.

Knowledge is our means of production.
Woven into nature, ourselves, and our creations,
knowledge converts energy of the universe
into nature's wealth,
our wealth.
Much knowledge cannot be grasped
or held by monopoly,
be it public,
be it private.
Yes, socialism and capitalism differ.
No, we cannot have one without the other.
Our "means of production" is nothing less
than the knowledge which helps us maintain our species
—to survive.
Art and government belong in the lists of means.
What do we say in saying:
own them?
socialize them?
capitalize them?
We say only confusion.

‖

We should not mistake capital in an ephemeral industry for fundamental means of wealth production in our biosphere, species, and society. The petroleum industry combines petroleum (the energy knowledge it contains) with some capital (knowledge of how to find and process the petroleum). In a socialist country, government should operate this industry; under capitalism, a private company should. However, neither government nor private company can hope to take total ownership of our knowledge of how to use petroleum, including cruising downtown on Saturday night, beauty treatments, ... , pyrotechnics. Without society's use-knowledge, the petroleum industry produces no wealth. Humankind had wealth before the petroleum industry came into existence, and it should have more wealth after the industry fades in importance. A change in human knowledge made petroleum part of wealth. Further changes in knowledge can reduce its importance long before we exhaust the supply.

In this dynamic world we often raise questions as to who has and who should have the knowledge and the tools for change. The answers vary and may often confuse. Yet, fixed commitments to the traditional dichotomy between socialism and capitalism are worse. They compound confusion by pretending it does not exist.

28. Center and Edge

Socialism is the part of our economy in which our group distributes wealth according to values set communally. The process need not resemble consensus. The group may be active, passive, enthusiastic or bitter about delegating the distribution authority. Marketplace is the part of our economy in which we trade wealth according to values set by many people responding to supply and demand. Marketplace and socialism are distinct, but they help each other. Neither can replace the other. Neither can manage all of our capital. Socialism flourishes at our centers. Marketplace flourishes at our edges.

> *I hold these truths to be self-evident:*
> *Lake without shore is no lake;*
> *Lake-shore without lake is no shore.*

‖

In the zealots for socialism and the zealots for marketplace we have among us those whose views of economics are akin to saying: "My sacred lake shall have no shore;" and " my sacred shore shall have no lake."

Socialism and marketplace differ because of the societal space each occupies. We organize ourselves into overlapping groups. Each group has edges. At the edge it must interact with other groups and with its natural environment. Among other options at the edge, it may trade wealth according to rules of marketplace. Each group has a center. There it distributes wealth through a communal system that may include tradition and rules.

A typical family administers its resources communally. Children do not buy food from parents. Tradition says that children have a right to food, clothing and shelter. They have a right to be wealth consumers. Wealth producers in the family pool their resources to support the family. Authority figures establish and enforce codes of conduct. The family typically divides tasks among members according to tradition, authority and volunteerism. There may be great competition among family members, but ultimately the traditional communal authority structure decides who wins and who loses.

At its edges, a family competes and cooperates in a larger society with other families, individuals, and institutions. As a producer in modern marketplace economies, the family may compete against the rest of the community for jobs and sales. As consumer, the family probably enters the marketplace to bargain for the goods and services that it does not produce internally. The marketplace, not a communal system of decision and authority, decides the prices the family will pay and receive.

Family members find themselves at the center of other institutions—a club, a church, a government, a corporation, a school, a clan, a tribe. At these higher-level centers interactions are communal, because larger institutions also practice socialism. Many participate in marketplace at their edges as well.

29. From Amoeba to Market

We can, I suppose,
think of a chemical reaction
at the nucleus of a tiny amoeba
as but one reaction
in
 a
chain
..that
..extends
from this amoebic nucleus
to the farthest reach of the biosphere.
We should rather, I propose,
notice this amoeba has a surface,
its working edge,
where universe divides in two;
where biosphere divides into
what-is-this-amoeba
---and------
what-this-amoeba-is-not.

Where Universe Divides In Two

Early creatures had centers and edges. Evolution gave them ways to move energy around the center, from edge to center and from center to edge. Communal biological wealth in the amoeba keeps its primitive power and control systems functioning. At its edges, the amoeba uses genetically coded knowledge to capture food. However, one amoeba has no control over the likelihood that food will be there to capture.

There is friction. One amoeba may find itself in competition for food with other amoeba, other creatures. Lacking the sophistication to sign treaties or to create a marketplace the amoeba relies on programmed gathering techniques to bring fuel into itself. At its edge the amoeba competes and cooperates with its surroundings to overcome friction, to transfer wealth from the larger biosphere into its one-cell body.

As life forms become more complex, so do the options for wealth transfer at edges, and for wealth distribution and recirculation at centers. Humankind has experimented with a variety of methods to organize its centers. At some scale we call these experiments family. At other scales we call them government. We have a propensity to form governments and never seem satisfied with our results.

As family met family, group met group, and nation met nation, we experimented with government-type solutions at our edges also. As alternative to war, raids, walls, and moats, the marketplace was such an experiment. Participants agreed upon marketplace rules in order to minimize risk to themselves and their goods. These rules and their enforcement were the government of the moment, the communal base for marketplace.

Fringe trading deals with new goods and services more readily than does communal tradition. We need only a willing buyer, a willing seller, a place to bargain, and security for the individuals, their goods and services. Traditional distribution may create successful investment, but that same tradition may be inadequate to guide the distribution of new types and quantities of wealth produced by the investment. With or without the sanction of traditional leaders, individuals might resort to edge-type trading to distribute these new goods and services.

Some governments came to approve of this internal trading, but also took control -setting and enforcing the rules. Thus, government became communal sponsor of free enterprise. Internal edge-economies became important. Marketplace society was born.

30. Corporation: Delegated Socialism

I ask you to think back to when the Soviet Union had the archetypal large communist government and good-old Ford Motor Company was the archetypal large corporation. Though they symbolized competing ideologies, these large organizations were in many ways similar.

When it dealt with countries in a peaceful interchange of wealth, the Soviet Union competed and cooperated in many of the same ways that Ford Motor Company competed and cooperated with other businesses. At their edges they both competed with corporations, individuals, partnerships, and governments that sold similar goods and services, that wanted the same resources, and that recruited similar personnel. The marketplace determined the price they would pay or receive. At their edges, then, looking outward, capitalist corporation and communist country practice marketplace.

When Ford Motor Company looked inward to its own resources, it acted—and still acts—very much like the old Soviet Union. Executives allocated resources under policy set by a central committee or board of directors. Executives and board lead in the directions that they believe will most benefit their constituents—whether shareholder or proletariat. At their centers, looking inward, both capitalist corporation and communist country practice socialism.

The United States of America operates the same way. When we look inward, we set communal rules for wealth distribution, be they liberal or conservative. Our country allows more internal marketplaces than either the good-old Soviet Union or good-old Ford Motor Company. However, "allows" is the important word. Communally we create, enforce, and administer rules that foster and allow private marketplaces. Without communal enterprise private enterprise could not work.

Ford Motor Company would not allow the level of internal property and private enterprise allowed in U.S. America, or even the old Soviet Union. The very purpose for incorporation is to socialize risk among investors. Corporations require public charters. They seek such charters because a charter will protect their investors. Individual liability is allowed by charter to be socialized into corporate liability. Individuals can profit and recover their investment severalfold, but their liability for loss cannot exceed onefold —their investment.

Corporate employees experience socialism. Until hired, a job candidate at Ford Motor Company is an independent competitor-cooperator for the corporation to bargain with at its edge. Corporate resources and individual talents get weighed and bargained. Ford Motor Company can reject the applicant or make an offer. The applicant can accept, reject, or haggle over the offer. Once hired, however, she becomes an employee. Ford Motor Company tells her what to do and limits her discretion in making decisions. She is expected to work for the common good, which she hopes coincides with her own. Ford Motor Company pays her the same for Tuesday as for Wednesday, no matter her production on each day. Even if the company offers pay incentives or lavish bonuses based on her performance, higher officers of the corporation set the incentives. Ford Motor Company also supplies her work space and the tools of her trade. This would not be the case if somehow her employment were marketplace based rather than communally based. Her promotions would come from her own success at marketing her talents or products to several buyers. Her status today would not ensure tomorrow's employment.

In the corporation the employee is likely to strive to maintain her secure employment and climb the corporate ladder. Among those who wish to climb that ladder, competition is fierce, but promotions and salary increases come from corporate decision makers above. There are rules that govern job security and procedures which define the role of each worker. The corporate authority structure distributes corporate wealth among workers and share holders. Authority figures also decide whose ideas will carry the

81

greatest weight. Corporations tend to prefer employees who accept this system rather than collective bargaining or individualism. This top-down decision and wealth distribution structure mimics family tradition, socialist tradition.

None of this is bad. We should expect a private corporation, like any organization, to be socialist at its core. What is bad and misleading is to champion corporations as the epitome of free enterprise. They are not. They epitomize delegated socialism—a societal attempt to organize centers of socialism where free enterprise or national enterprise might do poorly. If we have reason to look for the epitome of free enterprise, we do better to look among industrious individuals, sole proprietors and partnerships that owe less of their existence to delegated socialism.

31. Marketplace or Socialism: a Dynamic Choice

If a good or service is to be traded in a marketplace, its supply must be limited. The good or service must be identifiable. It must be unique or divisible so that buyer and seller believe they know exactly what they are transferring. We cannot trade most wealth in a marketplace because we cannot easily identify or divide it, or because it is too abundant. Neither the oxygen produced by the world's flora nor the bile working in our livers is apt wealth for trading under most circumstances.

Some wealth that we can identify and divide does not make it into our marketplaces simply because our communal tradition says it does not belong there. We do find wood-burning heating stoves and firewood in the marketplace. A well-designed stove is a storehouse of functional and aesthetic knowledge. We can buy it and its fuel. Yet, we tend to give away the heat they produce.

My stove knows how to burn its firewood,
how to respond

to me who knows so little of what it knows,
to me who does not know how to make a stove.
My stove knows how to send smoke up its chimney
and warmth into my room.
Its warmth can please,
or it can save a half-frozen life.
Such is its success and popularity
that I could sell tickets to my stove's proximity.
But, I do not.
I share its warm knowledge freely
according to communal tradition
among family, neighbors, and kindred strangers.
Those whom my stove knows to please,
those whom my stove knows to save
give back nothing in trade
—except,
to carry forward in common tradition
what we and a stove
must in-common know.

My Common Stove

No scientific standard tells us which identifiable goods and
services should be owned and traded in the marketplace.
During the Civil War in the United States of America many
young men bought up their obligation to serve in the army.
Government had made this military obligation a commodity,
making such purchases perfectly legal. During later wars
many people found legal ways to avoid military service and
combat, but it became illegal and immoral to try to buy one's
way out. Military service had become a duty; it was no
longer a commodity.

Voting is a right in some countries, a duty in others.
Individuals possess this right or duty. They could sell it, but
law and tradition say we can neither sell nor buy a vote. Law
and tradition try to keep votes out of the marketplace.

Law and tradition make businessmen criminals and criminals
businessmen. Traders in alcohol went through these

metamorphoses when U.S. America entered and later left Prohibition—when alcohol sales were made illegal and then again legal.

Necessity and changing styles can move goods and services into and out of the marketplace. A corporation chooses a new combination of socialism and marketplace when it discontinues the internal manufacturing of certain components in favor of buying them in the marketplace from independent suppliers. At one time, perhaps, there was a real advantage to socializing the production within the corporation. Back in the late-twentieth century, when automobile makers Chrysler, General Motors, and Ford recognized some of the hints that they had become uncompetitive because of their aging technology and stodgy managers, they transferred some design and production from their center economies to their edge economies. There they hope to take advantage of the technological and price competitiveness of independent suppliers in a dynamic marketplace.

A family faces similar dynamic choices. After rearing children under loving socialism many parents request economic support from adult children who work for outside income but stay at home. While this is not a pure marketplace transaction, food and shelter, once distributed to the child under central traditions, now get distributed under a family-socialized form of edge economics. The change recognizes new conditions which call for a new economic mix. The added income helps the parents care for the younger children or enjoy a little luxury. The working child learns the pride and responsibility of adulthood, gaining some independence from his parents. It is as if the edges of a new household (like a new cell) begin to form before it separates from the old. If the adult child becomes sick and unable to work, however, his family will reverse itself without hesitation. "Don't pay room and board now. We are your family; we want to help you recover." They return to pure socialism.

The marketplace always has and always will be adjusted by communal decisions, especially when the public finds a market to be distorted from what is just or sensible. If modified law and tradition do not correct such distortions, correction may come from war, or revolution, or massive government spending. During the depression of the 1930s the marketplace behaved as if there were little wealth to invest. Nevertheless, under the New Deal of President Franklin D. Roosevelt the Congress of the United States of America made massive communal investments to fight poverty, and then even greater expenditures to fight World War II.

While choices are dynamic, our leaders often choose to be dogmatic. From a great war and a great depression much of the world (winners and losers) emerged wealthier than before. Perhaps, if our politicians were less dogmatic and our economists more instructive about government's communal intervention in the marketplace, we could avoid great depression and great war.

32. The Difference: How Outcome is Determined

Ideas, people, and products compete intensively—both within a corporation and within the markets where the corporation buys and sells. Competition is often mistakenly equated with marketplace and free enterprise. Socialism fosters and tolerates every bit as much competition as does marketplace. Where they differ is in how results of competition get determined. In a marketplace, price (or value-per-price) is the arbiter. In a socialist system, the group or its authority figures decide based on merit, not price. Merit is judged according to criteria socialized into the system.

We uncovered certain socialist acts
perpetrated to settle competitive dispute.

Our Central Committee named a general secretary in

. the old Soviet Union.
Our board of directors promoted a manager
. at good-old General Motors.
Our boss sent one among us
. to the workshop in Hawaii.
Our parents chose one among us
. to finish the three-layer cake.

We Uncovered Certain Socialist Acts

33. Ignorance and Confession

A society that emphasizes marketplace will tend not to see the socialism that makes marketplace possible. A socialist society will tend not to see the adaptive potential of marketplace. I believe what I see: we can have neither a socialist nor a capitalist world.

We cannot survive without socialism; it evolved with our species. It is at the center of everything living. Theoretically, we could have a world without marketplace; it is a relatively new product of evolution.

There are alternatives to marketplace in edge economics. To organize the exchange of wealth at their edges, groups can compete through war, thievery, conquest, or isolation; they can cooperate through peace, contracts, treaties, compatible but independent traditions, or through intergroup (international) law. Contracts, treaties, traditions, and intergroup law can control interchange. Thievery can also bring in wealth (such as when an immigrant smuggled the secrets of England's looms to the American continent, or when a Native American stole from another group the first horses for his own group). Conquest can bring in all the wealth of another group. Isolation can put up real or de facto walls that prohibit interchange. However, humankind's accelerating introduction of new forms of wealth makes marketplaces almost inevitable, even if illegal in an avowed communal state. We can foster the marketplace, as we do in avowed capitalist nations, but let us remember that fostering is a communal activity.

We can neither totally delegate to individuals nor totally socialize to a group the means to produce and maintain humankind's wealth. If the Soviet Union had believed that the state should own the means of production, it should not have gone into the international marketplace to buy computers and wheat. It should have bought Nebraska and IBM—people included. If U.S. America believes that people should individually own the means of production, it should eliminate the state of Nebraska and the IBM corporation because these are strictly communal structures.

I make these absurd proposals only to point out the absurdity of labeling political and economic rivalries as a dichotomy of capitalism versus socialism. The wealth production potential of the USA suffers when it ignores the fundamental dependence of marketplace on communal control over the policies of wealth distribution. Because in the last twenty years of the twentieth century marketplace was in political ascendance, so was our ignorance of our communal center. Socialism was descendent. During these twenty years, socialist leaders tended to confess that their nations suffer when they ignore the vitality, flexibility, and healthy complexity that marketplace can bring to an economy. As times change so will the loci of ignorance and confession.

34. Sellfish on Cheritable

Robert Reich argued that, in order to arrive at a consensus for progress, we must acknowledge that the social and economic components of our national well-being are linked powerfully together. We must search for the right mixture of socialism and marketplace. This search is basic to resolving problems of food production and distribution. Frances Moore Lappe and Joseph Collins of the Institute for Food and Development Policy have done an excellent job of pointing this out. While small, individually owned and operated farms tend to be the most productive, nations increasingly look to large, corporate farms (or in the case of some communist countries, communal farms). In poor countries with little industry, governments promote agricultural exports in exchange for foreign goods and services. In the process, self-sufficient farmers become undernourished wards of the state. In U.S. America the family farm seems endangered as more farmers go bankrupt and corporate farming expands. These are complex issues from which I will draw only a couple of points.

The socialized ethic that says everyone should have enough to eat comes in a heritage older than our own species. In our enthusiasm for marketplace, we should never forget that only more recently in the biosphere's evolution did we socialize the market into our culture. It is no more inconsistent for a nation that values the dynamics of its marketplaces to make sure that every citizen has enough to eat, than it is for a family to feed all of its members—even if some produce nothing for the marketplace.

Similarly, a nation should never forget that if it fosters either corporation or commune it fosters delegated socialism. There is nothing especially natural or inevitable about either. They exist through communal law and policy.

Where individual families produce enough food
. (for themselves and nation)
. and have not wrought havoc with the landscape, why substitute larger,
. less efficient,
. private corporations and state communes?
Since corporations are socialized creations of government,
. there would be nothing un-American
. in keeping them out of farming in the USA.
Since families are communes,
. there was nothing un-communistic
. in reverting to family farms
. in countries leaving the Soviet Union.

Of course,
. the complement too is true.
. Corporate,
. communal,
. and cooperative farms may be useful
. where family farms cannot achieve
. marketplace, social, and environmental goals.
There is irony in U.S. America
. (and nations that follow its example)
. where, in the name of free enterprise,
. the most free of enterprises
. (the productive family)
. gets displaced by less efficient groups
. that socialize risk and profit.
Communist countries displayed similar irony
. imposing communal structure
. that more mimics impersonal corporation
. than traditional communism of family.

‖

We can see human culture as an arena of continuously
overlapping organizations, ranging from small families up to
associations among nWations. Each organization is itself an
organ of others. Each has a communal core where it
determines how to distribute its wealth internally. Our arena
can also be seen as an intricate web of edges through which
organizations exchange wealth. At these edges, a
marketplace can facilitate interchange. Core socialism and
edge marketplace complement one another. The world and
its nations need not choose between them. The world and its
nations cannot choose between them. Rather, they can
organize themselves so that both socialism and marketplace
help to capture wealth nondestructively, distribute it fairly,
and recirculate it repeatedly.

For this, I coined a bad pun: *We sell fish on a cherry table.*

35. To Suggest a Framework

What I have worked
is just forgotten.
What I have wrought
is but forgot.

Ought To The Participle

I end Part One suggesting that we might re-subdivide the
science of economics. In the first Plum Local, I saw a need
for a new name to cover an expanded economics—perhaps,
"biosphere economics" or "world economics." Change in
name, however, is less important than change in substance.
The substantial changes I describe rely on distinctions
between abundance and scarcity, between centers and edges.
Here, I propose a framework to accommodate these
distinctions. I propose to unite similar areas of study within
economics and to help economics connect itself with the rest
of science through a redefined macroeconomics.

War and marketplace are moral near-opposites. However,
both belong under edge economics because they try to
bring wealth across the friction of the edge between
groups. A treaty does the same, however a treaty usually
belongs under central economics because it can form the
core of a new group. It is the center of a higher level
communal system.

Corruption and thievery are moral kin. But corruption is
central while thievery is edge. In corruption, a small
group at the center of a larger group finds its communal
way to subvert the larger communal wealth ethic. In
thievery, an individual or group imports wealth snatched
from others.

Socialism and marketplace fall in different divisions of
microeconomics, not because they are rivals, but because
they solve different problems in the way hub and rim of a
bicycle wheel solve different problems.

Macroeconomics *covers the broad scope of economics, its important connections with the rest of science, and the interrelationships among the centers and edges of microeconomics. It describes the work by nature and humankind that creates wealth. It describes the general rules of abundance and economic development. Some traditional macro and microeconomics belongs here, as does economic history. For the most part it is a new economic umbrella synthesized from other parts of science.*

Microeconomics *describes the work of wealth distribution and maintenance. It describes the special rules of scarcity. Microeconomics can be divided into two parts:*

Central Economics *covers socialism, law, tradition, treaty, peaceful cooperation, money, stock issues, internal borrowing, corruption, and like subjects. Most of traditional macroeconomics belongs here, together with studies that have not traditionally fallen under economics.*

Edge Economics *covers marketplace, trade, war, peaceful isolation, external borrowing, thievery, and like subjects. Most of traditional microeconomics belongs here, together with work from other social sciences.*

I trim my beard;
you notice my hair.
You cut your hair;
I ask, "since when the beard?"
On sight of a dirty bath,
together we throw out the baby.
My,
what able eyes for change
and lame brains for attribution
have you and I, sir,
and Little Red Ridinghood.

You and I, Sir

I prefer this framework, but I believe that names and frameworks can be both problem and solution. I seek a framework that works better, but we should avoid excess concentration on the classification of fields of study. After all, subdivisions are artificial; there is only one science. We need only make sure that economics be an integral part.

Part Two:
Community Investment
— topics 36 – 43—

36. Work for Economic Development.

Why get a new brick
if I can dig up an old one?
This brick has lane three places in twenty years.
I dug it up twice,
laid it down twice—
each time thinking the brick work was done—
each time pleased with my work.
Even today, it did not displease me,
but it will please me more over here.

Why Get A New Brick?

Economic development is the work of life. My father liked to quote his Hungarian father: "Nincsen munka, nincsen sonka;" "No work; no ham." Perhaps a less literal but better translation is: "No work, no live." Both my grandfather, who repaired street cars, and my father, who repaired railroad cars, were speaking of physical, often-unpleasant work, the purposeful expenditure of energy. They were speaking of labor, the human effort that economists say turns capital into productive wealth. Yet, "work" has another meaning that I am sure was familiar even to my grandfather who spoke English as a second language and only outside his home. If brakes he had repaired were to show signs of failure on the street car taking his family down a steep hill to a Sunday picnic, "work" would have had the other obvious meaning: "If these brakes don't work, we die."

Work is function, effect, and, in physics, the application of force over a distance. Work is not brute labor. It does things. Most good students work hard to learn, but excellent teachers and parents know that much is taught through play. Play is every bit as good as work—if it works. I look around at nature and see much of it at play, yet I do not hesitate to describe an interesting flower or phenomenon as "nature at work." Science, art, and the Sunday picnic are human nature at work. I like to think that they have done as much for our economic development as have Monday in the office, Tuesday in the factory, or Wednesday on the farm. If you think not, I hope you will at least agree that they do work for our economic development.

37. Organize for Economic Development.

Economist, please try to define
why our wealth (1) can grow and (2) decline.

Please Try To Define

Life had to learn how to work for itself. It had to learn to sustain and expand itself in its home, the biosphere. It now knows how to organize itself to capture and expend energy with purpose, and to produce results that function. As life expands, it learns more ways to organize. Life's library of workable knowledge sustains it during tranquillity and changes it during crisis. The library grows larger and more diverse, even as it loses some works—species and ecosystems that have disappeared.

Knowledge is the wealth of the biosphere, the product of good work and good fortune. It organizes biospheric development, human development, human culture. Biological knowledge encoded in genetic libraries sustains the complex processes that mold inanimate energy into animate energy and animate energy into complex patterns. Brain knowledge adds real-time, real-life learning to the library. Artifacts, artifactual knowledge, organize and maintain energy in ways that genes and minds cannot.

In ancient Egypt, farmers who knew little of engineering drew upon the greater engineering knowledge stored in irrigation structures, systems, and administration to increase food and fiber production from their work. As they learned which plants did best, they stored their improving knowledge in each generation of seeds—selecting seed from superior plants. The improved varieties were their living artifacts, storing knowledge for children who used that knowledge without relearning it. They simply planted seeds of the improved varieties. Brain knowledge also passed from one generation to the next through written and spoken words and diagrams. And, of course, this knowledgeable culture could not have survived if men and women had not known to reproduce themselves and nurture their offspring into adulthood. Complex biological, brain, and artifactual knowledge were both essence and cause of a productive culture—not perfect knowledge, not perfect culture, but alive and long-lived.

Economics should explain how our species organizes itself for work that produces and sustains well-being. It should explain why "prosperity" is not "well-being" unless it works to promote likelihood of survival in our species and our biosphere. In short, the economist should describe the rules of organization for economic development.

Wealth is our preoccupation—
it is not our invention.

Economic development is an organized invasion—
life invades the universe.

Do not try to create economic development.
Be, instead, a knowledgeable partner.
Practice stewardship and equity.
Learn more.
Share knowledge.
Know to organize to foster the biosphere.

Know that wealth can disappear—
species, ecosystems, and cultures become extinct.
Guard against
the natural disaster,
nuclear war,
and revolution
that can destroy knowledge—
leaving machines that no one can use,
books no one can read,
science no one can remember.

Guard against the disappearance of life,
for the biosphere knows how to become wealthier.

Prevent the destruction of our species,
for we can learn to partake of that increasing wealth.

Stop exploiting one another,
for then we can partake of more wealth more quickly.

Bring our understanding of the world into harmony
with the knowledge embedded in the universe.

Thus,
we shall act as consistent development partners—
in our human sphere
in our biosphere.

A More Knowledgeable Partner

38. There Are New People In Upper Forest.

I return to a diagram I used earlier to portray wealth
development in the biosphere. Our biosphere is the largest
organization of life, sitting astride part of the inanimate
universe, isolated from any other biospheres that might exist.
Energy flows into the biosphere; or, better, the biosphere
invades the inanimate universe and captures energy.
Organisms within the biosphere distribute and recirculate the

energy among themselves, thus sustaining and expanding the biosphere. Energy eventually flows or is cast back into the energy streams of the inanimate universe as biospheric waste products.

The biosphere is an organism. It's components are sub-organisms. However unruly we seem, we components organize by rules known to the biosphere. We may participate actively or passively in our organization; we may participate in ignorance or with knowledge; but we participate. Fate is uncertain. Biosphere may expand its share of universal energy or it may collapse into poverty or death —so too, our species. The process is essentially imperfect, allowing disorganization, reorganization, flawed reproduction, new forms and levels of organization. Such is the biosphere and its economic development as portrayed in the diagram.

Now instead of seeing the diagram as the biosphere at work in an inanimate universe, I will change scales and describe it as a developing culture in a place called Upper Forest. A pioneer family enters a well developed forest ecosystem whose energy (- - ->) flows from plant to animal and back, ever mixed with new inanimate energy from the sun. Used energy flows either into the inanimate atmosphere or into a neighboring ecosystem.

Upper Forest is a complex organism, containing millions of sub-organisms operating at hundreds of levels. Yet from the viewpoint of the pioneer family, Uppermost G, none of the wealth of the forest is human wealth until the family enters the forest to harvest its resources. The Uppermost G's remove trees, feed piglets, extract food and materials, build structures, and leave waste to rot or otherwise reënter the forest ecosystem.

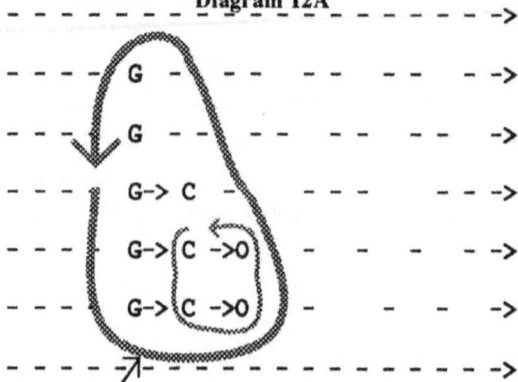

Diagram 12A

The biosphere and its layers of developed wealth.

The Uppermost G's are reasonably successful. Children are born. They increase consumption but eventually increase the family's capacity to capture and use the wealth of the forest. However, the G population expands even faster, because the Uppermost G's welcome another family of relatives, the Nextmost G's, who quickly take up the same lifestyle. An ever-more organized community forms as, over time, the Thirdmost, Fourthmost, and Bottommost G's move into farm the Upper Forest.

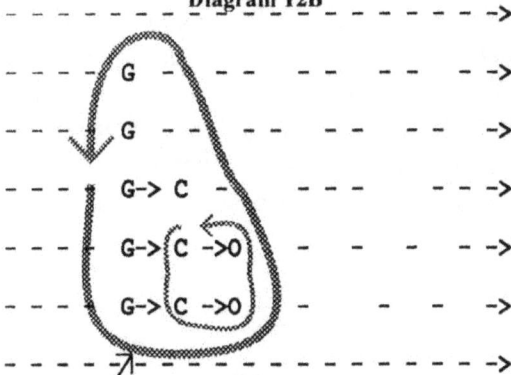

Diagram 12B

Upper Forest and its layers of community.

Some of the older children of these families see an opportunity to deepen this simple community and change from farming generalists to horse-and-buggy transportation specialists. Ever dependent on the farming output to pay them, they are proud to be the C'people who take goods and passengers to and from market and around the community. Eventually, others see the opportunity for leather working. The O'ers begin to craft useful goods from the hides of farm and wild animals harvested by the G's and C's and trade the products back to the G's and C's for food and transportation. Two O'ers and two C'people developed such a close working relationship that they formed the C-O Transport and Leather company.

The Upper Forest human community has expanded its farming activity five-fold from the time that the first

G'family established itself in the forest. However, community wealth has expanded more than five-fold. Due to the specializations of the C'people, O'ers, and C-O Co, much of what had gone directly to outside markets or had rotted quickly back into the forest stays in the community to enrich the lives of everyone from the Uppermost G's to the Bottommost O's. The Upper Forest Community sees that it has achieved significant economic development. These people might say their success comes from hard work. If Upper Forest Ecosystem had a voice it would likely protest. It did most of the work while the human residents just played around. As the impartial critic, I would try to calm the argument with, "So far so good. Whatever you all have done here together seems, for now, to have worked." Then, with a little diagrammatic cut-and-paste, I would point out that they are not alone.

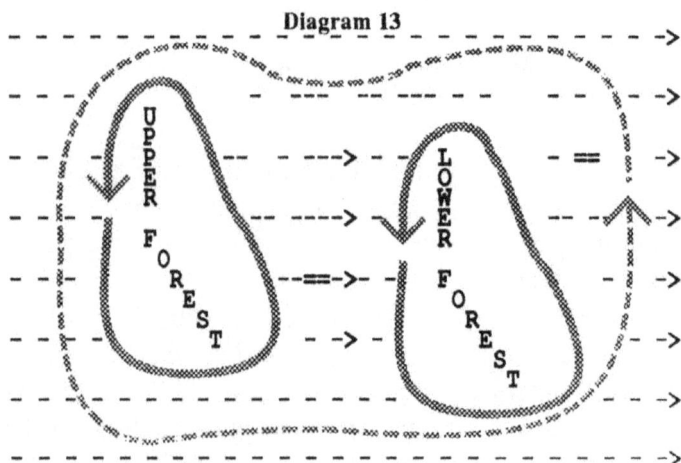

A Watershed Community.

Lower in the forest lies another community much like Upper Forest. Lower Forest inhabitants depend on Upper Forest because their river starts up there. Directly and indirectly the human community and ecosystem of Lower Forest depend on resources provided by Upper Forest. Even much of their valley soil has come over millennia from gradual erosion in the hills above. Waste products of Upper Forest community become energy inflow, wanted or unwanted, to Lower Forest.

Upper Forest community may for a while see the clearing of trees for agriculture on steep hills as economic development. Lower Forest community sees dramatic changes in the flows of water and sediment into their valley. Floods become more frequent. Siltation makes their irrigation structures and their potable water system unusable.

Lower Forest community may retaliate: close roads, take political action, resort to violence, or end the intercommunity trading that has been profitable to Upper Forest community. Perhaps both communities will find the wisdom to see that they interdepend. They are parts of a larger watershed community, a higher level organism. Each part sees economic development at the local community level and inward but should also see it from the local community outward to the watershed community.

39. Six Acts for Love of Wealth and Biosphere

We capture broadly—
> *as leaf captures sun,*
> *mill captures wind,*
> *and gatherer gathers grain.*

We distribute deeply—
> *as leaf sends oxygen,*
> *mill delivers flour,*
> *and parent feeds child*
> *that teacher educates.*

We recirculate densely—
> *as we bake for our miller,*

who rewards our harvester,
who buries our excess
to reward the roots
who will feed new leaves.
to grow another leaf.

We aid one another:
in capture,
in distribution,
in recirculation.
Mitochondrion helps newborn.
Child helps family.
Family helps village.
Village enhances ecosystem,
species,
culture.
In the best of worlds,
all enhance our biosphere.

from: Capture Broadly

As organisms grow wealth grows with them; communities and ecosystems grow. As organisms die wealth dies with them; species, nations, cultures disappear. We can consciously participate in disappearance or its prevention. Looking inward, we of this village and we of this species can capture, distribute and recirculate our wealth. Looking outward, we can aid biosphere and its parts to capture, distribute and recirculate their wealth—our wealth. We can act for love of our wealth while we act for love of our biosphere.

You of leg,
you of twig,
and you, of course, bacterium,
huddle broadly,
densely,
deeply.

Please capture geotherm and passing sun.
Please distribute.

Please recirculate.

> *For love of wealth*
> *and of biosphere one,*
> *huddle densely,*
> *deeply,*
> *broadly,*
> *   you of wing,*
> *   you of wheel,*
> *   and you, of course, bacterium.*

Quick Note To Roommates

40. Baby and Student, Investment and Balance

If we measure economic relationships among nations by subtracting imports of trade goods and services from their complementary exports, we produce a figure that says something about trade surplus or deficit. It is a humble figure that has a secure place in the basement of accounting houses. Unfortunately, we often invite this figure to stride out of its basement nook, mount a dark horse, and ride off self-importantly into media and politics. If we fix our attention on this figure as it parades by with its precedents and successors in impressive curves and columns, we miss most of what is important about exchange of wealth.

Much wealth flows across time and international boundaries in ways that do not fit trade figures. If we climb above the parade and broaden our view, we can see that nations form just one layer among complex layers of overlapping, wealth-exchanging communities. We can see that the trade routes connecting nations are but a few conduits in the biosphere's complex network of wealth exchange.

We will see imbalance when we focus on isolated conduits. For trade to happen, local, temporal, item-specific imbalances must exist. There must be vacancies in a

103

housing market; there must be imbalance in trade; there must be imbalance for there to be investment.

Notice, as baby and student come by.
Baby imports wealth
far in excess of export.
Student does more of same.
Do you see an obscure trade deficit
or an obvious good investment?

‖

Imports have always exceeded exports in our biosphere. Otherwise, biosphere could not have grown and prospered. Biosphere, our prototype for economic development, does export byproducts to the inanimate universe, but never as requisite to an import. A community with trade imbalance may be investing well in children, education, quality of life, and survival. A business can be responsible while borrowing to invest in its future. So too, a community invests responsibly in its future using trade imbalance to borrow necessary resources from neighboring communities which voluntarily foster their own profitable, responsible imbalance.

41. A Brief Note on Community Artifacts

We build artifacts: transportation and communications systems, schools, libraries, institutions and machinery. We call them public works, and support them by taxing ourselves to build them. As we move through an information age artifacts change. The library building will be less important than the means of access to its information. Safe and convenient routes to recreation facilities may become more important than routes to office or factory. As entrepreneurs have greater geographic flexibility in choosing locations for their operations, the civic, cultural, and governmental institutions that provide a healthy and stimulating environment for worker and family increase in importance. No matter the changes, we will continue to spend much time and energy to supply ourselves with public works --artifacts that sustain our community.

42. One Dimension We Do Not Like

The more that science comes to know,
the more that we come to share it,
delegate it,
distribute it,
organize it,
teach it,
learn it,
build it,
keep it,
the better and the deeper
we develop our community.

‖

Since we extract from nature the wood, food, fiber, space and energy to support our culture, we may see irreconcilable conflicts between our wish to increase wealth and the biosphere's need to flourish. Greater extraction by us means less biosphere for itself. That conclusion seems reasonable, but it is wrong. Increase in human wealth does not necessarily, or even usually, require that we rob the biosphere.

If we were a one-dimensional species that could increase wealth only by adding to our breadth, economic development and extraction would be bound tightly. We could capture one more unit of economic growth only if we add one more extracting unit to the perimeter of our broad field of extractors --facing into the biosphere and grabbing what we can. But this image does not fit reality. To be truly one-dimensional we could only increase wealth by increasing population. New families in the population would have to be just like all others, and independent of them. They would work at our perimeter just as older families do at our middle.

If we were but broad, one-dimension extractors,
we could have no specialized crafts,
no division of labor,
(for they require organization in depth)
no government,
no clan, nation, or fraternity.
This is not economic development as we see it.

‖

We would not enjoy one-dimensional economic development. We would never feel better off. Our species might become wealthier as it gains survival insurance --a chance flood or disease would be less likely to destroy us all. However

We would be no more wealthy than our grandparents,
no less wealthy than our grandchildren.
A lifetime of hard work would bring
neither greater reward to ourselves
nor greater security for our children.
This is not economic development as we like it.

‖

43. Community B and the Back Rub

Increased breadth in our community economy does not by itself make us feel more wealthy. However, increased depth and density can make us feel wealthier. To see this, imagine two isolated communities that have been stable, both in depth and breadth. Their populations hardly fluctuate; government and traditions are stable. They each have stable economies supported by the harvest of fruits from the forest, a few native crafts, and by entertainment that includes dance and dream interpretation. The two communities are identical, with one exception.

Community B recently discovered the art of back rub. The back rub has become a desired service in Community B. Several skilled back rubbers stay busy.

While Community B harvests no more food or timber than before, some individuals do harvest extra so they can give it to the back rubbers in exchange for a rub. The back rubbers have been freed of their need to harvest. Having those needs satisfied and time to spare, they can be induced to give a back rub if others offer food, a dream interpretation, a dance, a party. Gradually, the interchange of all goods and services in the community has increased. Dream interpreters do more. There are more dances. More people get invited to more parties. To accommodate the newly desired good, the community has increased the density and speed of its distribution and recirculation. It has grown in depth and density.

While extraction has not increased, people have become more efficient in their use of the resources they extract. This is partially due to skill improvements that come with increased specialization. But it is mostly due to a new incentive to do more with less, to ephemeralize. Ephemeralization has occurred in the use of fruits of the forest. One way to get a back rub is for specialists to carve ten wooden bowls and leave less waste from a branch that would yield only eight using traditional methods. These two extra bowls might buy a back rub, or a dream interpretation, or a dance. Thus, while extraction has not increased, the amount of energy bound into the wealth system of the community at any one moment has increased.

Residents of Community B feel wealthier. To the residents of Community A they look wealthier. In real economic terms they are wealthier. Even by conventional measurement, the economic development is obvious. The gross community product shows a marked increase. How? Gross community product is calculated by counting the same energy several times as it appears in different places and forms. This redundant counting correctly reflects increased wealth in the community.

Of course, events did not have to come out this way. The back rub might instead have overburdened social and political systems. The resulting havoc or revolution might have made the community fall backward or apart. Such is the risk of change. Economic development must be imperfect or it will stop. Yet, the Community-B experience, economic development in depth and density, does happen.

Part Three:
National Investment and Money
—topics 44-50—

44. A Better Money Legend

Who Created Money?

Money is one of humankind's great inventions. I was not around when it was invented, but I do think I know how it happened.

In the third or fourth grade I was taught: primitive people traded things for things; we needed a better means of exchange; we invented money.

What an image! These hapless primitive people dragging around elephants and chickens to exchange for coconuts, while their suave modern cousins sit under a tree sipping coconut milk assing a few coins back and forth. I have since heard many adults and children tell this same simple history.

My hair was turning white when I found something missing —Who?

Granted, trading chickens for coconuts could be cumbersome for the poor folks who walk to the marketplace, but poor folks do not create money.

Kings, queens, emperors, bankers, and empresses create money. They always have plenty of burly helpers to carry their chickens and coconuts.

Which one said: "My poor subjects are so overburdened. Trade is so bulky. I, nice person that I am, will give them a means of exchange."

Even if one leader were that smart and that good, why did cruel and dim despots continue the practice?

Listen to this story to find a better legend.

Let us imagine that first money was made of gold,
and it was issued by an empress. Before making her coins
the empress needed gold. Whether she stole it, got it through
in-kind taxes,inherited it,or had it mined in her own mines,
she had a storehouse of gold.Gold held great value in her
empire. She could buy armies, loyalty, roads, music, or
whatever.

Now, suppose her artisan made her a radical proposal:
"Your highness,you are so beautiful;I am so skilled. Let me
melt your gold and cast it into little disks carrying my
beautiful rendering of your beautiful face."

She feels flattered and tempted. But she says,"no," and puts
the artisan in the dungeon for treason. After all, some gold
would spill on the foundry floor. The artisan could hide away
bits of the gold—making him a rich and disloyal man. Many
laborers would be needed. The melting would use up many
trees. The dungeon was almost too little punishment.

Then there arises a great crisis in the empire. In the east a
horde of barbarians prepares to invade. In the west, three
princes want a good road to market. Without it they may
sever their lands from the empire. The empress summons her
advisers. They report:
   An army to defeat barbarians costs 10,000 pounds of
   gold. A good road over the mountains to the west costs
   10,000 pounds of gold. In the empress's storehouse are
   10,000 pounds of gold. The barbarian invasion by itself,
   or the secession of the princes by itself will bring
   destruction of the empire and death to the empress.

Despair sets upon the empress and her advisers—until she
remembers the artisan. "Free the artisan. Instruct him to
melt all my gold and mint me 20,000 gold coins carrying my
image. I proclaim, each coin has a value of one pound of
gold. Spend these coins to raise an army. Defeat the
barbarians. Build a road over the mountains to the west."

*In her selfish interest the empress spends coins. Her subjects
must accept them at twice their value in gold. Through this
great and official hoax she proposes to save her empire and
her life. To her own amazement, the hoax works. She builds
the road. The western provinces boom in productivity
as ever more commerce uses the new road. She defeats the
barbarians. In the eastern provinces, under peace and
stability, farmers and artisans produce as never before.*

*Once coerced to accept the coins, citizens now covet them.
With more goods and services to buy people welcome the
new means to buy them.*

*The secret cannot be confined to one empire. Kings, queens
and emperors imitate the great hoax. Some invest their coins
wisely and prosper. Their subjects revere and follow them.*

*Other rulers are less wise. They spend their diluted gold
coins on pleasures that drain wealth rather than nourish it.
Drowning in a sea of over-valued coins, their subjects resort
to trading of a-good-for-a-good, and they rise up to throw
out their worthless rulers.*

*Who created money? Someone powerful and selfish and a
little wise. Someone like the empress.*

Who Created Money

45. Money is Paper.

I believe the study of national investment would be easier
(and better) if we were to start with the assumption that
money became a means of exchange when created by selfish
rulers who spent it for their own good, and when their
spending (by accident or by design) proved beneficial to
many of the people under their rule. I do not guarantee the
accuracy of this history, but it illustrates why money is still
around—it works—and why money is essentially paper even
if minted from gold. No ruler ever paid the price of pressing

a stack of gold bars into coins if the total face value of the new coins was not significantly greater than the value of the gold bars that he melted down to make the coins. That greater face value gave a handsome and easy profit to the ruler. If he spent that profit wisely many of his countrymen lauded him as noble for the national prosperity he caused. If he spent it unwisely he had trouble keeping his crown. The paper profit is the true value of the coin. While gold content may have made new coins more acceptable, eventually the precious metal would be needed only to make it difficult and expensive to counterfeit. As money became more common and counterfeiting became easier to control, the precious metal content of money could go down until the metal could be replaced by paper—and now by the vapor of electronic bookkeeping occasionally reported on paper. Paper money is more portable than metal coin and has been difficult to counterfeit. Inevitably, simple "I.O.U."s written on plain paper, paper checks and other plain-paper documents, electronic recordings, and verbal promises could circulate as means of exchange—as money. When people believe that they can exchange paper for goods and services, and when their experience supports this belief, they gladly accept the paper, use it, and call it money.

46. Money and Other Biospheric Taxes

Once in the marketplace, money simplifies the exchange of goods and services. This is the obvious effect that we all see. Less obvious is the tax-effect. Money is a mobile tax. Once government becomes a minter of money it engages in monetary taxation—inducing people with newly minted money to play roles they would not otherwise have played in the provision of goods and services. Light, portable, magical money shifts wealth flexibly from person to person and place to place-reorganizing society as it flows. True taxation occurs whenever a minting government spends its money.

Any country can invest in good projects using its own money. Within the country venders must accept national currency in exchange for their good or service. They will tend to do so voluntarily because good projects create wealth which new money lets them share. Thus, as good investment follows good investment, money becomes the common way to exchange wealth in the marketplace.

Money is a magical invention, but not a freak. It is but one invention by humankind in a chain of magical biospheric inventions that allow a whole to organize itself to be more than the sum of its parts. Whether several micro-organisms exchange energy, a family allocates chores, a culture defines roles and traditions, or a government mints money, the net effect is for the group to tax itself—to allocate resources in ways that individuals cannot or will not.

47. Three Siblings: Loan, Stock and Money

Every time I use my credit card to buy something I issue a piece of real or symbolic paper—paper which embodies a loan secured only by my promise to pay. A bank gladly accepts my new paper. Its accountant calls my paper promise, an asset. With that new asset my lender bank can invest more than it could lend yesterday. Every time a business buys something using its line of credit it issues a piece of real or symbolic paper—paper that embodies a loan secured only by the business's promise to pay. The lender bank gladly accepts the new paper as a new asset. Today's new asset makes the bank worth more to buyer or investor than it was worth yesterday. The paper loan—floated on nothing but a promise—is an unsecured loan. This unsecured loan is a sibling of money.

Money has another sibling—a fraternal twin adopted by corporations. Common stock floated by corporations is the fraternal twin of money floated by national governments. Both are worthless paper with an initial value in the marketplace based only on speculation that proceeds from sale of the paper will be invested wisely.

Many corporate leaders and investors, and perhaps you, believe a national government should run like a business—claiming that a government is unbusinesslike if it spends more money than it collects. This dogma makes it difficult to see the family resemblance between stock, loans and money. Please prepare to suspend your belief for several paragraphs. If you hold this dogma the most I can ask is that you consider what you are about to read to be science fiction. Try to enjoy reading of a different world. Later, over a cup of your favorite drink, please ponder the possibility that this different world is our real world.

When good-old General Motors and I had good credit ratings we were allowed (often encouraged) to not balance our budgets—to paper the world with more credit than we can pay for immediately. Lender and investor wanted to hold our paper because they believed they would be well repaid for holding it.

Good-old General Motors issued paper called stock. Each new issue reorganized the investing public so that some of us gullible people played our assigned role and gave General Motors our money in exchange for the stock. If Jane had not been so induced she might have put a California hot tub on her back porch. Part of her wealth would have flowed to a small hot-tub company in California. Instead, it flowed to a giant in Michigan. Mary got laid off in California and moved to Michigan where her cousin, Sid, had just been hired for a new General Motors' project.

When good-old General Motors invested Jane's money poorly, it experienced inflation: Sid got laid off; Mary failed to find a job; Cadillac prices went up, or profits went down; the value of General Motors' common stock went down.

Jane and other stockholders took a risk for potential reward. Her risk and potential reward parallel those of a holder of national currency. If government invests well money holders will be well repaid.

Suppose good-old General Motors had required that anyone who wants to buy one share of its stock must pay for it with one share of Ford Motor Company stock. This may seem strange because General Motors normally asks for payment in money. However, if we suppose a time when one share of GM stock was equal to one share of Ford stock the stock-for-stock policy would be reasonable. It is as reasonable as holders of USA dollars exchanging their money for Canadian dollars of equal value. If General Motors had been trying to take over Ford it would have been quite reasonable to give Ford Stockholders new shares in General Motors in exchange for old shares in Ford. The newly issued General Motors stock would have been balanced by new corporate wealth-the assets of Ford. This is an understandable transaction with real balance.

Now, suppose good-old General Motors had been asked by its investors to achieve balance in the way advocates of balanced budgets say national governments should achieve balance. That is, General Motors had to receive one of its own shares before issuing a new one? Here is balance; nothing does balance nothing. Do you not agree that it is ridiculous balance? It would have been ridiculous to require that General Motors take in one share of its common stock for every new share that it issued. Such a requirement would invalidate accepted practice in corporate finance. It is equally ridiculous to ask the national government to take in one dollar for every one it issues.

A corporate stock issue should be balanced by an increase in corporate wealth coming from responsible investment of the proceeds from the stock issue. A national money issue should balance itself in the same way—through good investment. The public and private investment enabled by the expenditure of new money must create wealth to match the increase in money supply. To review:

Sibling one, Unsecured Loan

We borrowers must invest the wealth of our lenders in
productive ways that by choice or impossibility those
lenders will not invest directly. We trade unsecured, paper
promissory notes for a loan of investment resources. If we
do not invest those resources well, we will fail to pay off
loans, our credit rating will fall, our loans will lose value
(inflation), and our lenders may rise in revolt to take over
our remaining assets.

Sibling two, Common Stock

Corporate directors must invest the wealth of their
stockholders in productive ways that would be difficult or
impossible for individual stockholders to invest. The
corporation trades paper stock for investment resources. If
it does not invest those resources well its stock will lose
value (inflation), and its stockholders may rise in revolt to
install new directors for the corporation.

Sibling three, Money

Government must invest the wealth of its citizens in
productive ways that by choice or impossibility will not to
be private investment. National government trades paper
money for investment resources. If it does not invest those
resources well its money will lose value (inflation) and its
money holders may rise in revolt to take control of the
nation and its government.

48. Efficacy Limits

There are limits to sensible, productive paper investment.
They are efficacy limits. If some investment works but more
investment does not work, then more is too much. At some
point resources and society cannot respond efficiently to the
competing demands. At that point some investments fail.
They are good ideas that prove ineffective because they are
poorly timed. Efficacy limits are not dollar limits. Money is
sibling to unsecured loans and common stock. Like good
borrowers and good corporate stock issuers, our national

government will produce better budgets when it seeks to make good investments that are well timed. National government need not concern itself with the amount of money that it has on hand nor the amount expected to come in. It should rather consider what effect investment and non-investment will have on the wealth of the country and the international community of which it is a dependent.

The insidious implication of the balanced budget fantasy is that a national government would seem to never have to apologize for its expenditures if they do not exceed the amount of currency returning to its treasury:

"We ruined the country
and much of the rest of the biosphere,
but we never ran an unbalanced budget."

from: A War Rages

49. Banks Add Money and Investment.

Would you rather have a thousand dollars worth of stock or a thousand stocks worth of dollars? You might say, "That depends." Would you rather have a thousand potatoes worth of dollars, a thousand potatoes worth of stock, or a thousand stocks worth of other people's promissory notes? I suppose, "It still depends." With some creative logistics you can use any of these assets at the stock exchange to buy stock, at the bank to secure a loan, in your will to endow a university, or, in a pinch, most anywhere to trade for most anything. Why? Because there are secondary markets for potatoes, stock, loans and money. After the first transaction that puts them into the marketplace they can be sold again and again. The secondary markets for potatoes, stock, and loans are giant-but-tiny compared to the fluid markets for money. As the principal means of exchange, money is coveted by one side in most transactions. In modern economies we need a lot of money—so much that even free-spending governments can have trouble keeping up with the demand. They need assistance. While good-old General Motors would never allow anyone else to issue its stock (traders were stuck with the number of stock General Motors chose to issue), the demand for money is so great that most national governments let banks assist them in minting. The economy needs new money as enterprise brings forth new goods and services that the public wants. By choice or by default, governments in countries that emphasize the private marketplace let the banking system issue much of their money. Within the limits imposed on them by government, banks issue new money whenever they borrow from their depositors to issue new loans. If the banking system did not print more money there is no way that most bankers could cover their growing obligations to depositors—to return the deposits, with interest, on demand. Nor could borrowers pay their obligations to the banks.

Bank-printed money is easy to see if we consider the banking system as one bank. One day a man who is known to be reliable and productive takes 100 dollars out of his pocket and deposits them in the bank. A day later he borrows 80 dollars from the bank. The morning of day three he withdraws his initial deposit plus one day of earned interest. Now he has 180-plus real dollars in his pocket and an obligation to pay back his loan. He goes out and spends those dollars at stores that quickly send someone to the bank to deposit their earnings. By close of the business day the bank has assets of more than 180 dollars in deposits and more than $80 in good loans. As the cycle goes on, the amount of money in circulation and deposited in the bank continues to rise. Someone printed money, and, oh yes, during these three days government was on a holiday.

Having once put new money into circulation by lending out money that they have guaranteed to hold for depositors, a bank cannot withdraw that money from the marketplace even when investments turn sour. Borrowers have spent the money. It disappeared into the marketplace—eventually returning to the bank, but with no tag to identify it with the initial loan. Where projects fail borrowers never get enough money back from the marketplace to return what they owe to the bank. New money spent on these bad projects stays out in the marketplace searching for goods and services. Since the projects did not produce their promised economic growth, new money competes with old money for old goods and services—forcing up prices. The value of all money declines. That is inflation. Whether bank or government issues money to finance bad projects the effect is the same: inflation. Whether bank or government issues money to finance good projects the effect is the same: economic growth.

Despite the inflation risk, a banking system can serve its country well. A diverse investment structure that includes banks of various sizes in different locations can develop diverse investment strategies. Local banks can help a nation develop the complex organization at all levels that typifies a robust ecosystem. Because government spending should reflect priorities set by the nation, national government spending is the most direct and understandable way to infuse the private investment market with the money it needs. Centralized investment, however, even if intelligently made, even if made by a giant private bank, tends to over simplify the needs and opportunities of a complex society. A complex system of local and regional banks is a complementary money pump to national investment. It fills investment niches that simpler national strategies cannot find.

50. Investment Debate is Better Debate.

Neither a national government nor its agent banks can spend wealth that the nation does not have. Citizens expect reasonable balance between new money invested and new wealth produced. They expect monetary balance. It comes from prudent national investment, just as corporate stock balance comes from prudent corporate investment, and ecological balance comes from nature's tendency to invest prudently in itself. These are dynamic balances. Credits can more than balance debits as biosphere, species, nation, and family take dominion over a greater expanse of the inanimate universe.

We should hope to do better
than balance our budget.
We should hope to do better
than debate how best to limit spending.
We can save inferior debate
for the day the sun turns off
and the biosphere dies.

ǁ

A nation that understands the economics of money and communal investment will debate whether proposed investments contribute to the long-term health of society and its economy. Liberals might point to the profitability of social programs. Conservatives might argue strongly for hardware. In U.S. America some debaters will point to the boom in computers and related paraphernalia as wealth that followed investment in the space program; others to the black and brown faces in media and business as economic return from investments in civil rights and social programs; others to the Works Progress Administration under the New Deal as argument for a nation putting its labor force to work when the marketplace fails to do so; others to examples of peace work bringing higher return than war work; and many to the peace and productivity of those people who choose to invest their efforts to enhance the natural environment that they rent from future generations. These will be a difficult debates. For a change, however, they will be meaningful.

Productive investment debate happens within a successful nation that has prudent government knowingly financed by paper money. Growth in this nation's supply of paper money balances itself through growth in national wealth—a collective wealth that incorporates local, international, and biospheric wealth.

Investment debate is better debate.

Part Four:
National Finance; Debt and Taxes
—topics 51-78—

51. Against Foreign Loans

Poor countries continue to go into debt to other countries and international banks. They must repay their debt using one of the world's hard currencies—their own, considered too soft. Too often, some lucky citizens in the debtor nations manage to hoard hard currency that comes in through loans, while their governments hoard only the debt. Such problems hobble government and exacerbate the dichotomy between rich and poor in debtor nations, but damage is not restricted to debtors. The world suffers waste. Immense resources, including human effort and intellectual talent, pour into the system that supports international debt. Assume for a moment I can convince you that the system is invalid. Imagine the benefits we would reap if we divert this waste of resources into productive endeavor—endeavor that improves investment within nations and the trading of resources among nations.

As 1987's September turned to October. The members of the International Monetary Fund and World Bank met in Washington D.C. confronted by continuing crises in world finances. They started their meetings with little hope that they would devise a solution to the international debt problem. Their pessimism proved justified. By Christmas, banks in U.S. America had begun to write off large parts of their international loan portfolios as bad debt. If international debt gets less popular attention today than in 1987, perhaps the novelty is gone. International debt is an old problem that still stymies financial leaders.

Certainly some international debt comes from ill-conceived loans on bad projects, but other debt piles up from projects that were good-but-slow-to-prove-it. Money leant to enhance ecological, educational and social systems, for example, may return manifold profit in increased wealth, but the profit may come too slowly to cover the interest on the loan. Other projects may return profits quickly to society in general but too slowly to the marketplace. The market, not reflecting true benefit and detriment, declares the project a failure. Internal, national investment can accept some inflation as tradeoff for long-term and non-market benefits from good-but-slow projects. Instead of paying off the debt directly, the nation lets its money lose a little value—achieving the same end. An international loan, however, requires prompt payment to avoid geometrically escalating debt. Any country that embarks on good-but-slow projects by borrowing foreign money invites default—unnecessary default.

Much in life is unnecessary but quite acceptable, even enjoyable. International debt is unnecessary, but also harmful. I reach the same conclusion if the lender is a private bank or a multinational institution such as the World Bank or the Interamerican Development Bank. I reach the same conclusion even for projects that promise to be good-and-fast. I am against foreign loans.

52. The Unnecessary Burden of Foreign Debt

In Guatemala they have a way to carry heavy loads — primitive, simple, and efficient—with one or two loops of rope closed by a strip of leather or cloth. The strip fits the top of the forehead. The rope engages the load which may be a quintal of maiz, a household of furniture, or pots stacked five feet high by five feet wide. The body tilts forward to balance the load. You see men with such loads up steep mountain roads where you do not know how long ago your bus passed a town or a house, or how many rises and turns till the next. They learn the art with the smile and spurts of a six-year-old.
They continue when wrinkles and posture would seem to say "no." They also have a way to weed corn —primitive, simple,

*and efficient— with a machete and a stick. The stick is
narrowest where it fits the hand. A subordinate branch
makes a hook at the other end. The stick, though well-chosen
and well-fashioned, may be left near the field to be recovered
tomorrow or fashioned anew from a living fence. It gathers
and supports grass and weeds as the machete cuts them off
at the ground. The stroke begins high and vertical. An agile
wrist and low body quickly take it down to the horizontal.
The movement fits in the confines of the corn. The worker's
short stature seems to answer a plea from his bent body
which would refuse to straighten if its hips were higher.*

*Few men who have daily weeded corn fields and carried
heavy loads live to be old in years. When they do, their
bodies still need the food which bending and carrying
provide. I see one such man every day on the road to El
Rosario. His constant companions are his machete and a
boy. I do not know where they live. I usually do not see
where they work;
I see them on the road between. They come early to race
their day's work against sun's rise to oppressive heat. But,
they come later than most. They may leave the highway with
other workers, but on this four to eight kilometer trek they are
soon far behind. The old man does not move fast. The boy—
perhaps his grandson, nephew, or adopted friend—is in no
hurry to go faster. His pace is youthful but always a little
behind the man. The boy has someone to follow. The old man
—he is never last.*

*I cannot know if it is the years of work with the machete
or the carrying of heavy loads that has bent that back and
humped those shoulders. But, as the man walks by me now,
his eyes focus on the ground two paces in front of his feet.
Back is tilted; head is bent—as if to carry a load of corn. I
think he can no longer carry heavy loads. The machete in
one hand hangs as burden enough. Such is my judgment. Yet,
they seem to have a way to defy such judgment —primitive,
simple, and efficient—with one or two loops of rope closed*

*by a strip of leather or cloth where wrinkles and posture
would seem to say "no."*

from: Heavy Loads

Good governments finance what they hope will be good
projects and services—whatever the source of funds. Good
projects and services are successful. They benefit society in
excess of the resources that go into them. Even developing
countries have experts with sophisticated training to evaluate
the likelihood of success. Lenders who make loans to
developing countries presumably have people with similar
training to make the same evaluation. Experts do not know
for certain whether a project will succeed or fail, but success
is their goal.

International debt is a burden to the developing countries that
carry it. I argue that it is unnecessary; the international
lending structure need not exist. To illustrate and simplify
my argument, I will name Guatemala as a representative
developing country and the United States of America as a
representative foreign lender.

Guatemala is a small country with a fairly large international
debt. It is a wonderful and troubled country. In 1998 we hope
that recent peace will hold. In March of 1986 I held high
hopes for a new democratically elected government under
president Venicio Cerezo which had launched an Economic
and Social Reorganization Plan. Within 18 months, however,
Cerezo's government acknowledged that the plan had failed.
A key factor in the failure seems to have been under
spending. Of the amount that it had budgeted for public
investment, the Guatemalan government spent only about
one third. The limited spending failed to stimulate the
economy. Faced with ever decreasing living standards and
continuing civil war, the government abandoned its first plan
and began a new effort on August 1, 1987, The National
Reorganization Plan. The new plan was less ambitious and at
least as unsuccessful. The first Guatemalan plan had
borrowed some ideas from the New Deal and U.S. America's
recovery from the Great Depression. Public projects would

124

help get things moving. However, national coffers went dry. International loans did not make up the difference.

When we hear of the suffering that afflicts many developing nations that have large foreign debt, we should pause to consider that in accumulating this debt their governments probably tried to follow norms of international finance.

Guatemala followed today's norms for responsible finance when it refused to undertake beneficial projects for which there was no money. The public projects that Guatemala proposed but did not undertake were designed by dedicated and well educated planners. It is reasonable to suppose that some of the projects could have stimulated the economy and reduced poverty. Cerezo's government refused to undertake projects for which it had no money. Under conventional wisdom, in failing to act the Guatemalan government acted responsibly. Did it act correctly? If they were good projects, no. Should international lenders have filled the gap and financed these good projects with loans? No. An international loan is never more economically feasible than is internal financing, and international debt is never less burdensome than internal debt.

53. Buy Good Projects.

A good project is a good investment;
a bad project is a bad investment—
> *whether building life,*
> *species,*
> *family,*
> *association,*
> *or nation.*

A good investment pays
yet bears no debt.
What would the biosphere owe
(and to whom?)
for the inter-galactic loans

that financed its development?
Did the Gulf of Mexico
ever repay its debt to the Mississippi River?
Who could tell Mahatma Gandhi
that he had repaid all investments in his life?
Good Projects Are Good Investments

A good project, a good program, raises the value of the
nation. It can lead to deflation even if paid for with newly
printed money. A bad project or program causes inflation
even when paid for with money in hand. If the government
of the United States of America happens to have a million
excess dollars, to waste them on a failure would damage the
economy and the value of the dollar. Even if the Guatemalan
government happens to have four million excess quetzales,
wasting them on a failure would damage both the value of
the quetzal and the Guatemalan economy. A money-in-hand
project puts no new money into circulation, it does use and
alter resources and produce inflation as the old amount of
money chases after reduced resources.

If failure is to be paid for with an international loan it will
cause even more inflation. In order to pay off the loan the
borrowing nation has to export resources equal in value to
those invested in the project plus an additional amount to
cover interest. The net reduction or alteration of resources is
greater than if the same failure were paid for with money-in-
hand; the value of national currency falls more.

Now, consider a good project. The Guatemalan government
has plans for a four million quetzal project. All analyses
indicate the social, moral, and environmental results will
range from acceptable to beneficial. Analyses also show that
the project should increase the wealth of the nation well
beyond the four million quetzal investment. Unfortunately,
the treasury does not have four million quetzales on hand.

This project should proceed. It will be deflationary even if
Guatemala prints new money to pay for it. If analyses are
close to correct, this good project raises the value of the
nation beyond the current market value of the new money to

be spent on it. That is deflation. If Guatemala were to borrow foreign money to pay for the project, the result might also be deflationary. However, the interest payments added to the project cost will lower net national worth below that achieved through the issuance of new money for the same project.

54. Money Works Better Than Debt.

Let's look at two graphs of possible outcomes from three hypothetical projects: A, B, and C. Project A is a resounding success. B is marginally successful. Project C is a failure. The graphs show the net national benefit or net national loss from each project. Where there is a net benefit, the wealth of the nation increases and the net-benefit line on the graph ends above the centerline. That is a successful project. Almost all projects will start out creating a net loss because resources are spent before benefits return. The project that never creates a net benefit is a failure.

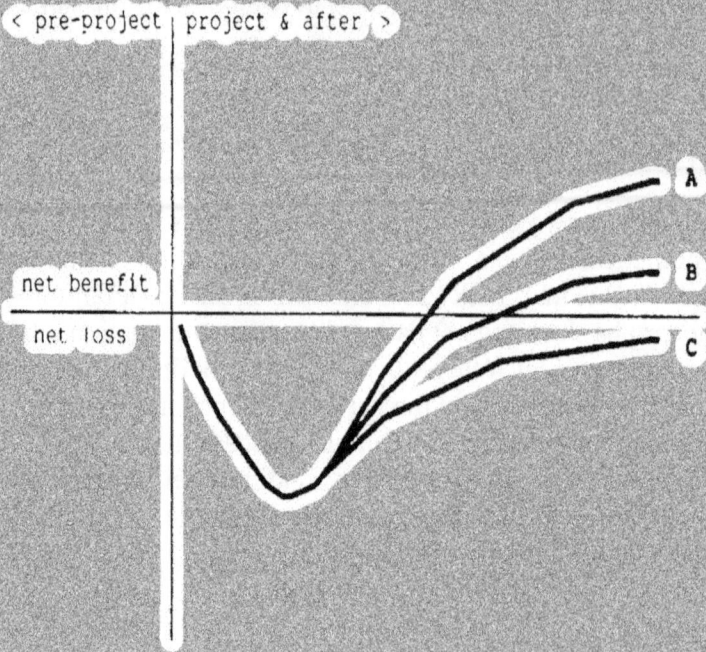

Graph 1
Three Projects Funded Without International Loans

< pre-project | project & after >

net benefit

net loss

A

B

C

Project A : SUCCESS
After an initial period of great costs and little benefit, the project creates substantial benefits that eventually far outweigh the early losses.

Project B : MARGINAL SUCCESS
After an initial period of great costs and little benefit, the project creates benefits that slightly outweigh the early losses..

Project C : FAILURE
This project does produce some long-term benefits, but they never outweigh the real costs. It creates a net loss in wealth.

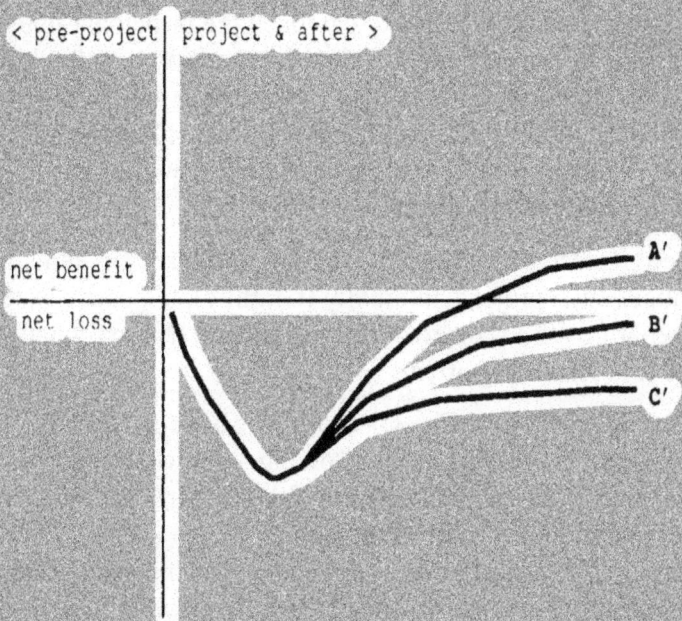

Graph 2

Three Projects Funded With International Loans

< pre-project | project & after >

net benefit

net loss

A'

B'

C'

Project A' : MARGINAL SUCCESS
The successful project is still successful, but now it is a marginal success. Subtracted from the net benefits of project A are the national resources that have been consumed or exported to pay for the interest on the international loan.

Project B' : FAILURE
What was a marginally successful project without an international loan becomes a failure. Benefits do not accumulate fast enough to cover the debt burden.

Project C' : MAJOR FAILURE
This project is a failure even without an international loan. Borrowing internationall to a for it onl worsens the results.

Graph 1 assumes the three projects are paid for internally. Graph 2 assumes an international loan finances the same projects. The two projects that bring net benefit without the loan (Graph 1) bring fewer benefits when financed by an international lender. Project B changes from a marginally successful project to a failure—producing a net loss. C is a failure in Graph 1 but worse in Graph 2. Why?

129

If U.S. America decides to lend U.S. dollars for a Guatemalan project, then Guatemala will have to pay back the principal plus interest (Graph 2). Guatemala will draw upon the project's benefits to pay back the principal—the one-time cost in resources of the project. Any benefits left over, less the amount needed to pay interest, are the project's net benefit or net loss. The interest is the price charged by the USA to Guatemala for the privilege of borrowing. The added charge reduces Guatemala's net benefit.

When, instead of seeking an international loan, Guatemala finances such projects internally (Graph 1), the entire excess of benefit over cost accru es to Guatemala. No extra national wealth is lost or exported.

Under internal financing, good projects are economically successful. Under external borrowing some very good projects are economically successful, but extra wealth must be exported to pay interest on the loan. They are less successful, less profitable, than projects financed internally. Projects that could be marginally beneficial when financed internally will likely become failures when financed with an international loan. For projects destined to fail, an international loan can turn mere failure into disaster.

55. Don't Let a False Analogy Fool You.

> *I did not enjoy the story,*
> One opines;
> > *it is much too artificial.*

> *So it is,*
> Two reminds;
> > *but such is the real world.*

> *I suppose it is,*
> One resigns;
> > *I suppose it is.*

A Dialogue

I think most of us consider borrowing money to be serious business; we do not do it if we have in hand the money or resources to undertake our project; we borrow when we need money for good projects. A developing country would seemingly pay for its projects itself if it has the money to do so, borrow if it does not. As any individual, does not a country borrow because it needs the money? If it cannot collect the money at home it needs foreign money, right? No, wrong.

Clearly, the Guatemalan government borrows from foreign lenders because it believes it needs the foreign money, just as I believe I need a bank's money when I ask for a loan. But herein lies the great mistake—the false analogy behind loans to national governments. It is falsely assumed that the principles of private borrowing can be applied to borrowing by governments of sovereign nations that caretake their own currencies. This error built the international debt crisis.

When I borrow from a bank—in an unsecured loan—paper passes in two directions. I hand the bank a signed paper that is my loan agreement. In return, I get paper money. I try to prosper. By prospering I can pay off the loan, maintain a good credit rating, and borrow more if needed. By managing well my financial affairs, I also manage the value of the paper that I issued to the bank, my loan agreement.

I control the value to the bank of my loan by the skill with which I manage my financial affairs, however the bank has almost no control over the value of the money that it gives to me. While the bank tries to prosper and manage well its financial affairs, it does not manage the value of the paper dollars that it passes out in loans. The effect on the value of money that comes from the actions of one bank is insignificant. Money fluctuates in value due to forces that neither bank nor borrower can manage. Our roles are not symmetrical. While we borrowers manage the value of the paper we issue, moneylenders cannot.

An international, intergovernmental loan is different: both borrower and lender manage the value of the paper they issue. When Guatemala borrows one million dollars from the USA its representative signs a piece of paper called a loan agreement and, in return, receives other paper called dollars. Each paper dollar gives Guatemala a claim to some resources in U.S. America. Although dollars may be used for transactions outside of the USA, their fundamental value lies in a guarantee by the government of the United States of America that this money can buy a million dollars worth of goods and services in the USA. The amount that each dollar can buy is not guaranteed; the right to participate in the USA marketplace is. During the period in which Guatemala uses the borrowed dollars, the government of the United States of America protects the value of the dollars as best it can. It manages the total number of dollars in circulation as well as the general flow of the USA economy. It tries to maintain a healthy dollar.

The loan agreement signed by Guatemala gives the USA claim to some of the resources of Guatemala. The fundamental value of the loan agreement lies in a guarantee by the government of Guatemala that it will forgo use of somewhat more than one million dollars worth of its future resources to repay the loan plus interest. The government of Guatemala protects the value of the loan agreement as best it can. It controls the number of loans that the country has outstanding as well as the general flow of the Guatemalan economy. It tries to maintain a healthy loan agreement.

This international loan is symmetrical. Paper dollars buy a paper loan. A paper loan buys paper dollars. U.S. America hands Guatemala paper dollars signed by its treasurer and continues to manage the dollar's value. Guatemala hands U.S. America a paper loan agreement signed by a comparable government official and continues to manage the loan's value. In economic terms, the roles of the USA as lender and Guatemala as the borrower are essentially the same.

Timing is one incidental difference, but not as significant as it might seem. Presumably the outflow of dollars would be an immediate drain on the USA, whereas the drain on Guatemalan resources gets delayed until project benefits flow. The timing of benefits should accordingly be just the opposite—early in Guatemala, delayed in the USA. However, both economies are likely to make short and long-term adjustments that dampen the effects of timing. Dollars that leave as a foreign loan do not necessarily come home quickly to the USA to be cashed in. Some will circulate for years internationally before returning to extract goods and services from the national economy. The loan agreement itself is a long-term national asset that will tend to uphold the short-term value of the dollar. A rapid drain on the economy of the USA is unlikely. In Guatemala, a loan agreement may affect domestic resource allocation long before the payments are due. Guatemala should gain in the short-term as it takes in money from the lender. However, the short-term drain felt by Guatemalans may be as great or greater than that experienced by citizens of the USA. Anticipating the debt payments, their government will tend to favor investments in export production, thereby altering local consumption. The citizenry of Guatemala may experience this extraction of customary goods and services from their economy almost immediately.

Don't let a false analogy fool you. Borrowing by national governments is not like borrowing by you and me.

56. Poor Countries Do Not Need Rich Money.

One said: The Guatemalan government does not seem to have extra quetzales.
Can it lend them to itself
or, for that matter, to anybody else?

said Two: Yes, the government in the United States of America does not seem to have extra dollars.
(Its accounting sheets seem to show negatives
where extra dollars are supposed to be.)

Yet it lends new dollars to other nations
and invests them in its own.
It also uses them to wage war.

One said: I know such irresponsible practice is long since
condemned.
Have not those United States long since prospered?

said Two: Yes. Meanwhile poor countries long since work
to avoid such irresponsible practice,
and those poor countries long since remain poor.

One said: So, if the United States of America
lends and profits,
invests and profits
with money that they are supposed not to have,
might Guatemala try to do the same?

said Two: Yes, and it should.

||

If you can accept that the system for lending to sovereign
nations is invalid there is yet another practical question: do
not poor countries need dollars or some other strong
currency to participate in the international marketplace?
Politics, tradition, and organizations such as the World Bank
and the International Monetary Fund can prevent countries
from using their currency easily in the world marketplace,
but there is no economic justification. The beauty of
marketplace is that it adjusts prices to allow trading in
products of different value. Imagine a farmers' market in
which tomatoes from many farmers—varying in quality and
kind—cannot be traded except by using as currency the
prized tomatoes of a few powerful farmers. This is
unnecessary and unacceptable.

Guatemala, Brazil, Mexico, the USA and any other country
can finance good projects with their own currencies. Within
the country, venders of goods and services must accept the
national currency without question. Such is the history of

money. For off-the-shelf international purchases, project managers can buy what they need using local money converted to the foreign seller's preferred money at the current exchange rate. For international orders and long-term contracts, project managers and sellers can agree to a price tied to some standard such as dollars, marks, francs, yen, or a composite index. They can still make payments in national currency in an amount adjusted according to that standard.

Wealthy as well as developing nations will experience setbacks that create prejudice against their currency in world markets. A crisis in Guatemala may cause most of the rest of the world to stop for awhile accepting the quetzal. Guatemala might flood the market with quetzales by making too many international purchases in too little time. The value of the quetzal could drop dramatically. However, trade will resume. The marketplace can establish an efficient exchange rate. If the USA or an international organization wishes to help, it could offer in-kind aid or political support. It should never lend money. A loan will reinforce prejudice against the borrower's currency. Why should we accept a troubled country's money when we know the country has just received more familiar currency from a rich country?

I saw them working in a field one day.
He was swinging his machete.
I only looked for a moment,
probably because I didn't really want to believe it's possible.
He swung his machete all that day and many days since—his friend working beside him.
I would like to think
that the boy does a little more than his share,
but it's probably just the opposite.

They walk home looking no different than when they came.
The boy does not require conversation of his old companion.

The first time I passed,
the sound of a gringo-ish "a dios!"
caused him to turn his head to reply.

Now, salutations to me
are the same as to the other familiar voices on the road
—uttered quietly in the rhythm of the resigned pace.
On occasion, I have passed without saying anything
—an irrational form of sympathy.
But he passed giving no notice to me,
maybe a little relieved that his attention was not diverted
—his attention on the road just two paces away.
However, why should I walk by him in sympathy
when his companion walks with him in respect.
It takes a great man
to have worked that hard for that long.
Maybe, in reality,
he still can carry a good-sized load on his back.
And the boy too is looking just two paces ahead
—at that man he wants to emulate.

After all, you see him every day
on the road to work in El Rosario.
At least he was there yesterday.

Passing

57. *Weak Argument for Reverse Foreign Debt*

Despite some superficial differences, loans between national governments are reversible. I have tried to argue myself out of this conclusion, but I cannot. I find it difficulty to distinguish borrower from the lender in international loans to national governments—that is in economic terms. Though common wisdom and common prejudice see them as very different, borrower is lender; lender is borrower. If you have trouble seeing the similarity, try to imagine a loan in the reverse direction.

I have a weak argument for reverse loans to support a troubled country. In a reverse loan, the helping country accepts a loan from the troubled one in the troubled country's currency. The loans have no true economic justification.

However, they might have financial justification and could make a great item of conversation to help you survive a dull cocktail party.

A reverse loan could grease international financial machinery for a troubled nation. The reverse-loan agreement gives the troubled lender an immediate asset with which to guarantee international purchases. Also, troubled currency that flows out as loan must come home to be cashed in.

If international lending were reversed, richer borrowing nations would have to buy goods and services from the poorer lending nation using the lender's troubled currency. Since rich nations tend to influence trends and styles, their purchases from a poor nation might foster international interest in the poor nation's goods and services. At least, international markets would have to grow accustomed to dealing in the troubled country's currency.

How would a reverse loan be set up? Instead of U.S. America lending Guatemala one million dollars, Guatemala could lend U.S. America an equal value in quetzales. In return for the loan, the USA will sign an agreement with Guatemala in which it promises that after it spends the quetzales it will forego domestic use of enough USA resources to round up the quetzales it will need to pay back the loan plus interest.

Once it receives borrowed quetzales, the USA can spend them to buy the same coffee, sesame seed, beef, and ornamental plants that Guatemala would have had to export to pay off a loan had it borrowed from the USA. Alternately, the USA could exchange the quetzales for Japanese radios so that Japan could buy the sesame seed.

So, Guatemala can finance the same project as lender that it could as borrower. The quetzales that it lends out are not lost. They will return directly or indirectly to Guatemala as the USA spends them. The loan agreement is an asset. It can serve as collateral to finance purchases or to issue new quetzales pending repayment by the USA of those that were lent out.

My strong argument is that a loan from Guatemala to the USA is fundamentally the same as a loan from the USA to Guatemala. Therefore, each is useless in economic terms. If Guatemala could just as well lend quetzales as borrow dollars it might as well lend the quetzales to itself.

My weak argument is that if international markets are sticky due to bad habits and irrational prejudice a reverse loan in quetzales from Guatemala to the USA might inject the needed grease. With well greased wheels, a caravan of reverse loans will parade brilliantly compared to the overheated squeak-buckets going in the traditional direction.

58. Former Debtor Nations

When we abandon international loans we will see changes outside the walls of treasury departments of national governments. Projects that international loans would finance today may differ when financed locally. Some projects favored and proposed by international lenders will die because the target nation does not forsee enough benefit. Other projects will be designed differently. Local project managers are likely to use and develop local goods and services to replace more expensive foreign options mandated or favored by international lenders. Purchasing departments will make foreign suppliers compete in a free market—rather than in a restricted market that favors suppliers from the lending nations. Conscientious governments in developing nations will feel even greater responsibility for their nations' futures. They will not hesitate to undertake projects that are likely to produce and to distribute justly more wealth than they consume. They will see that newly printed and well invested money out performs newly signed foreign loans.

In setting fiscal policy, developing nations will be subject to foreign political and economic pressure, but perhaps the pressure will be little more than that exerted among developed nations. Officers of foreign lending institutions and the International Monetary Fund will lose their extraordinary influence over the internal policies of the former debtor nations.

"Former debtor nations," I have not suggested what can be done with the present international debt. It exists, built upon false principles and good intentions. I have no magic answer. Diplomacy, rather than economics, must eliminate existing debt. As to the future, in order to promote peaceful paths to mutual prosperity among nations we must eliminate both the practice and the machinery of foreign monetary debt. In its place we can hope for constructive cooperation among former debtor and lender nations who act with discipline and responsibility.

As child and parent the word, discipline, intrigued me. One moment it seems to mean a steady rigor by which an individual or group works through a challenge. The athlete who sticks to a rigorous training routine has discipline. So does the scholar who works methodically for years to uncover the mysteries of genetic inheritance, as does the former smoker working through a successful withdrawal. They have discipline. Calling it, self-discipline, is redundant. Directly related to this meaning is discipline as a profession. The discipline of the athlete is high jump, of the scholar, molecular biology. This discipline is also a verb. The microbiologist disciplines herself when she enters a rigorous experiment.

Another moment, discipline seems to mean punishment, structure, rigor imposed upon the unwilling. Parents who confine their children to their rooms on Saturday night are said to discipline their children. The department head who docks the pay of a scholar who did not follow rules for the use of the photocopy machine is said to discipline her staff. As a noun: "Children and scholars need discipline from their superiors or they will be irresponsible."

So, I chuckle when someone says, "There is a lack of discipline here," and everyone nods the head in agreement. They can agree about lack of discipline, yet they might have profound disagreement if pressed to elaborate their views.

Self financing issues no license for irresponsibility. Poor countries need discipline as much as rich countries. There is no special reason that discipline for poor countries must be imposed by the International Monetary Fund and international lenders. How do we prevent developing countries from acting irresponsibly if we let them print money rather than borrow it? Unfortunately, we cannot, with certainty, prevent countries from acting irresponsibly. Responsibility must be learned from the pain and reward that come after investment. We should not expect a country to borrow money it does not need and risk inflation it does not want so that it might acquire discipline.

To invest in the well being of a nation
requires good judgment.
To invest poorly
is to risk disaster.
Nations that invest well
do improve quality of life.
Those that invest poorly
do suffer.
Therein lies the lesson in responsibility.
The Lesson in Responsibility

59. Tax is in Our Roles.

Taxation is our tendency
to cooperate toward abundance.
It is the role of roles,
the sustenance of organization,
the carver of niches.

||

Few economic proposals attract more interest in U.S America than strategies to balance the federal budget. If you fear what will happen if the federal government spends more than it takes in through taxes, your fear probably comes from your acceptance of an intuitive equation:

Spend = Income

As spokesman for both the fear and the equation, President Reagan complained in his 1987 State of the Union address that over the past half-century the United States of America had outspent its means. For many people his complaint portrayed well the problem facing the country. To others, his words sounded like a cover-up for the national debt that burgeoned during Reagan's years as president. For me, Reagan's complaint reinforced my doubt in the validity of the problem as popularly stated. After all, despite some downs and ups, the country prospered during those fifty years. Some disadvantaged groups fell to further disadvantage. Yet, income, defined as growth in national wealth, increased.

Paradoxical.
Fifty years of lousy national management,
fifty years of deficit
between revenue and spending,
bring fifty years of unprecedented wealth.

Conventional explanations:
(1) We finance deficit,
sell the deficit to our citizens.
(2) We exploit
environment,
minorities,
other countries,
to make up the difference,
(3) Private enterprise makes up
for the failings of government.
Attractive and defensible explanations!

Yet, now a quandary.

Does bad management make a country wealthy
  if it   borrows from its citizens,
  fosters private enterprise,
  exploits poorer countries,
    weaker citizens,
    and a defenseless biosphere?

I offer two other explanations:
(4)   SPEND has nothing to do with INCOME
    when we correctly account
    national government's attempt
    to tax its own currency
    back to itself
    in order to spend it again,

(5)   SPEND always equals INCOME
    when INCOME refers to LEGITIMATE TAX.

‖

There are at least four legitimate taxes: tax in law or custom, tax in money, tax in kind, and tax in price. Each type offers advantages but is often interchangeable with the others.

Legitimate taxes start with communal policy; call it government policy. The taxes take effect when society responds. Society adjusts its pattern of role playing, creating a new distribution of wealth. Singers, carpenters, and bakers ply their trades in new ways and places. They change trades, change the way they ply their old trades, produce different products and services, change their consumption, move to new towns, go back to school. They alter their roles in society. In short, they pay their taxes. They are well taxed if their changed roles create new wealth and protect old wealth. A society that is well taxed has it's budget in balance.

60. Tax in Law And Custom

We tax ourselves in law and custom,
rule and ceremony;

role and symphony.
As best we can,
we tend to our interaction.

||

Each of our formal laws, each of our less formal customs channels our actions. Each is a tax upon us.

"Thou shalt be available for jury duty."

"Thou shalt put x% butterfat in our ice cream."

"We shall send our children to school."

"Men should wear ties to church."

"Our employees shall not work on the 4th of July."

Be it law or custom, each of these rules for participation in society taxes citizenry by inducing or demanding individuals to allocate their resources—that is, to play new roles or to keep old ones. The resources collected and distributed have, in effect, been collected by society's tax collectors and spent by its administrators—mostly common citizens, but society's collectors and administrators none the less.

In U.S. America on July 4th, Independence Day, we get broadly taxed by our law and custom. The nation accomplishes the role playing it desires. Most people go to the beach or to their back yards on the 4th of July rather than to work. Human and material resources get taxed from one activity—the typical work day—to another—celebration. The nation presumes that the holiday role playing is more beneficial than costly, so it takes resources away from a day's production in widgets and allocates them to a day's worth of celebration.

61. Tax in Money

Our government offers negotiable currency
to induce us to play our roles—
> *to provide goods and*
> *to provide services*

to support communal efforts.
In short,
our government mints and spends our money.

Most of us were taught that money is a means of exchange. Yes, once in the marketplace, it does simplify the exchange of goods and services. However, when first spent by government it is not a substitute for goods and services, but rather a mobile tax. Wherever our government chooses to spend it, our new money changes how we play our societal roles and allocate our resources. Each new dollar spent by our government reduces the market share controlled by old dollars—it reduces the percentage of marketplace goods and services that people holding old dollars can acquire. This is taxation, just as it would be if the government came to the door and physically collected the goods and services. Physical collection is unnecessary because our individual selfish acts, in pursuit of the dollars, collect the goods and services and take them to new places in the economy. Congress could, for example, decide to buy the July 4th holiday, compensating employers in cash. That would be a tax in money. Dollars spent by our government would attract employers and employees to a 4th-of-July holiday.

Most national governments mint and caretake a national currency. When they spend new money they tax their nations; when they spend new money they create income. Thus, the common balanced budget formula, **"Spend = Income"** is a tautology. It says only:

"Spend = Spend"
"Income = Income"

62. Tax in Kind

In kind, we give to our government

144

A group or its government might collect food to feed communal workers, fuel to heat communal buildings, and books to use in communal schools. Most groups do not print or mint money—family, United Nations, city, state, Girl Scouts, bridge club. They do sometimes collect from members the goods and services that the group needs. In doing so, they collect in-kind taxes.

We can see the July 4th holiday as a tax in kind, since employers must put up the time-costs of a paid holiday. The government of the United States of America wants most employees to have one day off from work to celebrate the anniversary of the nation's independence. If we see the employers as giving those days to the government which then awards them to the workers, we can say the employers have paid a tax in kind.

63. Whose Time, Whose Money?

In kind, we give to our government
*   some of our goods and our services,*
or, in kind, we may give to our government
*   strange money,*
*   once minted,*
*   once spent*
*   by a foreign government.*

||

The governing bodies of groups that do not mint and caretake money may tax their members in money. Even a national government can tax citizens in a foreign currency. Money minted by another government is a good, a mobile and flexible good that can be collected and converted into food, fuel, and books. The collected money is legitimate budgetary income. It is a tax in kind.

My insistence on the difference between minting and non-minting governments may seem puzzling, but it is the same difference that is obvious to a farmer who needs 120 hours of work to harvest his wheat. He knows that he can work twelve hours a day for several days to get the crop in. (He'll need another twelve hours each day to eat, sleep and attend to other important matters.) He can tax himself ten days of work to harvest the wheat. Unfortunately, this year, he needs to harvest the crop in one day because he expects a big storm. He knows that no matter how he uses his time, he cannot tax himself 120 hours in one day. He produces his own time, just as national governments produce their own money.

No amount
of self-taxation of self-production
creates more resources
to do the task at hand.

||

The farmer asks nine neighbors who have already harvested to give him tomorrow twelve hours of work each. Their organized 120 hours of effort bring enough time to the task to harvest the crop in one twenty-four-hour day. The farmer gets a tax in kind from nine neighbors. He taxes them in their time, not his time, and multiplies his resources for harvesting his crop.

The farmer knows he cannot get his own time back, call it income, and spend it again—knowing how foolish he would look if he pretended to do so. Governments that mint and caretake money must learn that they cannot get their own money back, call it income, and spend it again—no matter how responsible they look when they pretend to do so.

64. Tax in Price

Artificially,

our government raises
        (or lowers)
the price of a good or a service
so that
we change our patterns of purchase,
so that
        (for communal benefit)
we change our private investment.

∥

The U.S. American government does not forbid that we work
on the 4th of July, but rather requires that employers pay
extra to those of us who do work. That is a tax in price. We
raise the cost of labor on the holiday to induce employers to
give us the day off. When our government put a tariff on
certain electronic equipment imported from Japan, it raised
the cost to consumers in the USA. When it twisted the
collective arm of Japanese auto makers to put a quota on
exports to the USA Japanese manufacturers began to export
to us more expensive cars. The increased prices and changes
in products induced some of us to buy other products—
perhaps made in the USA by a Japanese firm, perhaps made
by a USA company in Mexico. Some of us may have
changed to the more expensive, more luxurious Japanese
autos, because, though it was more luxury than we needed,
we were willing to sacrifice in some other part of our budget
in order to have Japanese quality in our transportation. With
or without intent, our government induced these changes in
the way we play our roles as consumers. The tariff brought
money into the USA treasury, while the quota gave higher
per-item profit to the Japanese companies. Both tariff and
quota induced a reallocation of our resources in U.S.
America. Both were price taxation because changes in price
and product-for-price caused the reallocation. These
reallocated resources were the income (and loss) from the tax
in price. Any money that came into the USA treasury was
incidental by-product, not income.

147

65. *Monetary Recall: an illegitimate Tax*

Our Government tries, in vain,
to finance its actions
by recalling from us
—taking back from us—
a portion of the money
that it minted for us
and spent among us.

||

One government action is conspicuous for its absence from my list of legitimate taxes. The governmental action that we most often call "taxation" is an illegitimate (or pseudo) tax. I call it, monetary recall.

In ignorance, national governments recall money from their citizens, hoping to reduce the market wealth controlled by taxpayers and pass that investment power to the government. Since this pseudo taxation does not tax, it is no wonder that these national governments get confused when they try to balance their budgets. They try to balance real investments with pseudo taxes—fantasy taxes. A minting government can tax national resources by minting and spending money, but cannot get that money back—even though it believes and tries, and tries and tries again.

While few of us will kid ourselves into believing that the day we spend today is actually the day we spent yesterday, national governments continue to believe that the dollar that they recall today to spend tomorrow is the same one they issued yesterday. It is not because—in anticipation of the recall—wages and prices have risen to null the effect of the recall. Monetary recall generates no income for the minting government, rather it inflates the currency, decreasing its value.

66. Recall Seems to Tax, but Fails.

To an individual taxpayer, monetary recall seems to work. She can trade every dollar in her hands for a small percentage of the wealth traded in the marketplace. If the federal government recalls dollars from her—taking money out of her hands—she can now trade for less of that wealth. Each dollar she gives to the national government seems to reduce the wealth that she has to make purchases or investments for herself.

To the government official that receives her dollar the tax seems to work. He sees the recalled dollar as revenue. He sees the dollar that comes in as an opportunity for government to buy from the marketplace. The perceptions of both taxpayer and government official are understandable, but erroneous.

First, a superficial problem: taxation has costs. To debate, to legislate, and to receive, as well as to calculate, to send, and to avoid taxes we use resources. The recalled dollar buys less of what the government wants than it does of what the individual wants because part of the dollar erodes away during recall. In what amounts to circular bucket pouring, a little drips out with each pour.

Second, the fundamental problem: within a small margin of error, taxpayers know how much they will have to pay in taxes. They discount every dollar they receive, anticipating the return of part of each dollar to the government.

Into marketplace
goes the taxpayer
and with taxpayer
begins to bargain.
"After taxes I have nothing," he says.
"After taxes I will have nothing" she says.
After exaggeration and inflation,
taxpayer and taxpayer make a deal.

‖

Monetary recall is almost never a secret. It may be the biggest topic of conversation after weather and sports. People are smart enough to discount the face value of the money they receive to allow for the amount that they will have to return through the recall. Thus, it takes more discounted money to buy a refrigerator than it would take undiscounted money. Spread through the savings, lending, and investment system, the net effect of discounting is to force the economy to mint more money—to force inflation.

When a seller of chickens believes he should get the equivalent of one 75-cent cement block for each chicken, he will intuitively try to charge $1.00 if he knows 25 cents will eventually go for taxes. In turn, the cement block has a price of 75 cents because the cement block maker must charge that to get the fifty-six cents she wants from the sale of the block. The rest she sets aside in her mind for taxes.

This price-raising system is not perfect. Chicken sellers and block makers cannot raise prices without fear of loss of sales to competitors. However, the entire marketplace operates with discounted money. The value of a dollar is the value we give it. If together we discount its value in anticipation of monetary recall, together we push inflation. Buyers and sellers exchange goods and services for money in the marketplace with full knowledge that they have to pay taxes. All of the money collected by the Internal Revenue Service has already been discounted in value (prices have inflated). The value of each dollar reflects the net value that people believe they can receive from it. It incorporates a discount that reflects the average tax burden experienced by the population. If that average burden is about 25% a suit that would cost about $75 without recall taxes could cost about $100 with them.

When the price of oil went up rapidly in the early 1970's, other prices in the world's marketplaces tended to follow because petroleum users successfully raised their prices. Monetary recall has the same effect.

Income "taxes" raise the cost of my labor.

I raise the price of my labor because
I discount the value of my income
in anticipation of the taxes I know I must pay.
When I discount the value of dollars I receive,
I, in effect, demand that more dollars be printed.
I have become inflationary pressure.

We thought we had severely taxed ourselves
when, automatically,
we sent back some of our national currency.
Yet, we only taxed sincerity.
We thought we had paid
for what our government spent.
Yet, we only paid for what we did not earn..
For this is the rule:
Our automatic tax,
our national income tax,
puts more money into circulation,
and then tries as best it can to take it back.
If you doubt this rule,
please search for its exception
Search for:
the person,
the employer,
the employee,
the customer,
the idiot
who bargains and shops
ignoring that he knows
some of what comes in to him
goes right back out to his government.

II

67. Two Legitimate Recalls

Monetary recall is not without merit. It has two small but legitimate uses: price distortion and surprise extraction.

Monetary recall can distort prices in the marketplace in accordance with a government policy. A tax on beer favors the drinking of soda pop, for example. To discourage smoking, the USA initiated in 1998 significant increases in its tax on cigarettes. To induce capital investment it taxes ordinary income more than income from what we choose to call "capital gains." Such distortions are price taxation.

If imposed as a surprise, monetary recall can remove some currency from circulation. This surprise extraction is useful in times of rampant inflation. It works if government imposes the tax rapidly and one-time-only. Otherwise, the all-knowing marketplace will inflate itself to compensate. Sudden reverses in monetary recall also work for a short time. The tax cuts of the 1980's in U.S. America did put great quantities of unearned income into the pockets of the wealthy who were able to keep some of the inflation that the previous higher level of monetary recall had added to their income.

So, our national government can use recall carefully to remove money from circulation or to distort prices. It should, not, however, keep a general system of recall (such as the national income tax) just to be able to selectively distort prices and suprisingly extract money. Nor should it treat the recalled currency as income. This money is a by-product, an incidental possession, much like moonshine whiskey confiscated from illegal stills. We can burn this money and moonshine for heat, but we cannot use them to balance a national budget.

68. I Challenge The Doubter.

I challenge someone to prove that prosperity comes with budgets balanced by revenue from recall; that economic disaster comes from budgets not balanced by recall.

In U.S. America
unbalanced budgets
(during World War II and the space race)
appear to have caused unprecedented prosperity
(for most of us).
Neither Ford Motor Company nor I balance our budgets.
We paper the world
with more credit than we can pay for immediately,
Yet, nobody requires that we recall one paper
before we issue another.
So, somebody show us why
we should hold national government
to a different standard.

‖

69. Yes, Inflation is Real.

No matter how we do our national accounting, inflation is real. A national budget runs a real deficit if national expenditures fail to balance themselves through income: growth in wealth caused by the communal and private reinvestment of knowledge. What is today commonly labeled "deficit," or "unbalanced national budget," is an accounting fantasy spun from unreal definitions of income. These fantasy deficits get blamed for inflation. Though I dismiss deficit spun of fanasty accounting, I do not dismiss inflation. It is a real phenomenon. It can hurt us. It can also tell us how successful has been our investment.

Inflation may indicate that communal investors made bad investments. A project that fails can cause inflation even if paid for with cash in hand. While the cash-in-hand project creates no new money, it does redistribute, use up, or alter resources. When the project fails, the old amount of money now chases after fewer available goods and services. Inflation will haunt a nation that achieves its fantasy balances year after year but fails to produce and protect wealth with the meager investments that it does make. Frugality at the expense of wisdom will decrease wealth.

Inflation may indicate that the marketplace does not accurately reflect real wealth formation and loss, that successful investments have not produced salable goods and services fast enough to match the expanding currency circulating in the marketplace. Some benefits and costs do not reflect themselves in the markets or are slow to do so. Just as money should never be mistaken for wealth, the marketplace should never be mistaken as the only conduit for wealth, and monetary inflation should never be taken as the sole measure of successful investment. The low rate of inflation that modern marketplace countries tend to experience even in the best of times may reflect this divergence between total benefit and marketplace benefit.

While inflation never indicates that there has been too little monetary recall or too much fantasy deficit, it is an important concern for managers of money-based economies. We had best avoid inflation. We avoid inflation best when we invest well.

70. Consider the Cost of a Holiday.

Consider alternate views of a 4th-of-July holiday. Whether caused by law, by spending money, in-kind contribution, or price manipulation, a 4th-of-July holiday has virtually the same economic impact. The cost to the economy is the same, and the benefits are the same.

Ignoring for a couple of paragraphs the spending alternative, each of the other alternatives can cause its own monetary impact. Whether created by law, in-kind contribution, or price manipulation, the holiday affects the productivity of most businesses in the country and creates new businesses catering to the holiday itself. The marketplace reflects and facilitates these effects.

The net result might be negative—the holiday reduces national production, causes inflation. How? Since our chosen alternatives do not directly change the supply of money, if production of goods and services falls, each dollar could buy less than it could were there no holiday. Prices would rise as the same number of dollars chase fewer goods and services. Perhaps the holiday brings real benefits, such as reaffirmation of democracy and communal effort, but they do not reflect themselves in the marketplace.

The holiday might, however, be deflationary. Workers might respond to the goodwill and relaxation with increased production that more than makes up for the lost day. The 4th-of-July industry that caters to the celebrating public is a bonus. As a result, the marketplace would see an increased supply of goods and services to buy without any reciprocal increase in money to buy them. The market benefits outweigh the market costs; prices fall.

Now, let's consider the other tax alternative for creating a holiday—government buys the 4th-of-July holiday from employers. The pure economic impact of the holiday would tend to be the same, the costs and benefits the same. However, the financial-monetary-impacts would differ. In the worst case, if production of goods and services falls the resulting inflation would be greater. Even more money would chase after fewer goods and services. In the best case, the holiday and post-holiday increase in goods and services would be greater than the pre-holiday value of the money invested. That would produce some deflation.

Monetary taxation carries added risk, but it must also carry added potential benefit. Otherwise, only our stupidity could explain our continued fascination with the use of money to finance government. Once a monetary system is in place, monetary taxation may be the simplest, most practical way to achieve a goal. A bit more inflationary pressure might be perfectly acceptable if the non-monetary benefits are high and if it is impractical to set up price taxation, in-kind taxation, or taxation by law.

A tax in money offers potential benefits the other alternatives do not. In our monetized societies, the private economy generates a huge demand for money. As the economy grows money supply must grow. If money supply does not change with economic growth money becomes rare. Prices go down —which is nice for those who hold the money—until there is so little money that it is not useful as a common means of exchange. To prevent collapse of the monetary system, the caretaker of money must find ways to meet the demand for money. Government spending is an understandable, direct, and valid way to infuse the private investment market with money it needs. Thus, money spent to buy a 4th-of-July holiday can both finance the holiday and help fill a general need for money. Because needed money is demanded money, and because it enters at the 4th-of-July, the economy goes to the 4th-of-July to get it. This "going there to get it" is the essence of monetary taxation.

One way or another, by one tax or another, if we want a 4th-of-July holiday, if we think it is good for us, we will tax ourselves to get it.

71. Let Us Find Balance.

There is a true rule of balance.
An organization,
biological or political,
tends to survive
if self-organized and self-taxed
to be productive and resilient.
It balances its budget
with the wealth
that productive organs return
from the energy spent

to create and maintain them.
So, yes,
a government cannot spend
what it does not have.
However, what it must have
is access to the wealth
in the organs of its economy.
It has such access
whenever it fixes a policy,
or passes a law,
or, even,
when it spends newly printed money.

Let us end monetary recall
and the fantasy that it is a tax
that balances our budgets.

Let us find balance
investing communally
in the growth of knowledge
within ourselves
within our biosphere.

Let us recycle the effort of the intelligent people
trapped in non-production
calculating
collecting
debating
and avoiding payments
under a futile system of monetary recall.

Again, let us remember
that war is the age-old medicine
to counter peacetime fantasies.

from: A War Rages

72. When More Failure is Better

My conjectures on the fantasy pursuit of balanced federal
budgets were already well developed in 1987 when they got
a boost from an unexpected source. President Reagan gave
his State of the Union address to begin his last full year in
office. Measured by his speech, despite a collapse in the
stock market during 1987, the economy was in an especially
good state. Yet, he had entered office seven years earlier
saying two things had to be done to have a sound fiscal
policy:

(1) balance the budget; and

(2) reduce government spending

(thereby eliminating new deficits and reducing the old
deficit). Measured by his own goals he had presided over the
greatest fiscal failure of all time. Federal government grew in
absolute terms and shrank little if any relative to the total
economy. Its deficit would triple during his presidency.

This trend was apparent for each of his seven years. Yet,
each year he and his supporters repeated their two big goals
and then lauded the economic progress of the country.

I heard nobody ask the obvious question:
If years of unprecedented failure bring good,
would not more failure bring better?

I heard nobody voice the obvious:
Mr. President,
Your goals must be irrelevant.

||

73. Goods, Services, and Promises

I argue that it is wrong for a government which mints money
to recall this money to cover its expenses. It follows from
that argument that borrowing to make up for insufficient
recall is wrong because it is unnecessary. I have a hunch,

however, that you will hesitate before accepting the futility
of monetary recall. Also, I think borrowing is an interesting
topic. So, I choose to come at the balanced budget question
from the borrowing side.

My friends, too,
go where I want to go
to share what I am eager to share.
My enemies, too,
go where I want to go
to get what I am unwilling to give.

I argue that in economic terms the standard deficit
calculation is fantasy. In financial terms, however, it exists.
National government has borrowed a lot of money.
Necessary or not, the paper obligations occupy big spaces in
our marketplaces.

In our marketplaces people every day distribute and
redistribute the world's wealth. They buy goods and services
at prices derived from dynamics between traders. Depending
on our view of a transaction, we call one trader, buyer, and
the other, seller. Each wants to get as much as possible while
giving up as little as possible. They both have an incentive
to bargain and reach agreement on price because each is
threatened by the consequences of not getting the desired
commodity. Microeconomic theory bases itself on this
simple but dynamic relationship.

We might expect that in the perfect marketplace trading will
continue until everyone is happy with what she has and then
stop. This does not happen due to some obvious dynamics.
Goods and services get used and must be replaced. Tastes
and desires change; what we wanted yesterday we may wish
to trade away today. Old traders die; new ones arrive. Also,
even as we have world markets for many goods and services,
one marketplace cannot contain all the world's goods and
services. Even if it could, new goods and services keep
arriving while some old ones deteriorate, decompose, or
disappear. These dynamics do not bother microeconomics

much because researchers can usually design their analyses so that they can safely assume that the marketplace does not change enough to confuse their conclusions.

When venturing away from microeconomics, however, economists must concern themselves with a less obvious dynamic. Many of the newly arriving products and services have not arrived by camel or airplane from some distant market, but have merely appeared out of thin air. Goods and services are not just shuttling between markets. The goods and services available in the total of all marketplaces really grow. New goods and services regularly intrude in old marketplaces. These intruders from thin air are economic growth as manifested in the marketplace.

Even less obvious than these intruders is the trading in thin air itself. Traders deal not only in old commodities and new commodities but also in the mere prospect of new commodities. One gullible trader will give away real goods in exchange for the promise from a trustworthy trader that she will return next month with real goods in exchange. Using the resources she bought with a mere promise, the trustworthy one will go out and try to produce and maybe invent enough during the month to go back and pay her promise. The gullible trader who gives away the real goods for a mere promise knows that not only must he stand and wait, but he stands at risk. She may not come back and make the payment. It is reasonable for him to set a price that is above the going price for immediate trade in the marketplace. He adds some penalty for his wait and his risk.

While I describe this transaction as a sale, which it is, it is also a loan. In receiving immediate payment (in goods) and delaying her own payment, the trustworthy promiser has borrowed. The gullible one has lent her the value of his goods. By tacking a penalty to the price of the goods he delivered, the gullible trader has issued a loan which is to be repaid in principal (immediate market value) plus interest (penalty). Whether the exchange is in chickens or coins, it is a loan. Both parties do seem pleased with the promissory note; they entered into it freely.

Wherever buying and selling based on promises became popular, everyone could buy and sell more. They did not have to restrict themselves to the goods and services actually in the marketplace. They could speculate on the ones that would be there a month, a year, a day later. Merely by allowing this speculative activity, the marketplace grew. All the real goods and services were there together with the promised ones. Since both the real and the promised were being traded, the total value of the market was now greater than when only real goods and services were traded. The value grew even more when people not only entered into promissory notes but began to actually buy and sell the notes themselves. The man who sold real goods for a promise suddenly needs a real service. He cannot wait for the loan to be repaid so he buys the service using the promissory note given him earlier in trade for goods. When the trustworthy promiser returns she must pay the new holder of her promissory note. As this practice becomes understood and accepted, traders can deal in promises alone: "I'll give you these three one-month promissory notes for that one-year note you have."

Thus, one simple marketplace
expands into the once simple marketplace
where next to traditional good and old service
come the novel good and new service,
newly invented,
newly imported,
newly produced.
Then come
the promises of more to come,
and finally,
midst them all,
we find the market
for promises of promises,

for the some-day goods
and yet-to-be services,
to be sold
for the price of expectation
—for risk and promise
of what this marketplace may yet become.

‖

74. Money for Promises

Wherever money is a common medium of exchange, market
expansion creates a need for more money. If money becomes
scarce it ceases to work well as common currency. People
will prefer to trade goods and services directly, rather than to
round up the right amount of money. To prevent scarcity,
new money is needed not only to buy new goods and
services but also to buy new promises from those who say
they will bring goods and services to the marketplace. New
money is also needed to buy old promises from those who
bought them but now wish to trade them (at a discount) into
immediate cash. ("Discount" is another name for "interest"—
except that the buyer charges it to the seller when it is the
buyer who feels risk and delay.)

Money should get to the places in the economy where it is
wanted and needed. From where should it come? Ultimately,
it should come from the government that caretakes it, but
how? As the interpreters of public need, government officials
can get money where they believe it is needed simply by
spending it there. Usually, however, the thriving private
marketplace must attract more than government spending to
get enough money to cover expansion. To get the needed
money the marketplace, through its traders, attaches interest
to acceptance of a promise. New money must supplement
old money so that buyer can buy today's delivered goods and
services and seller can cancel the interest charged to her
earlier promise of their delivery. When promises are fulfilled

and interest paid, the marketplace has attracted the extra money it needs.

75. Deflation is Peril.

When Economists see that the borrower will pay more for getting what she wants one month early, and the lender will let her have it provided she compensates him for the month that he will not have access to her payment, they say that time has value. The interest tacked onto the loan by the lender compensates him for that value. The interest paid by the borrower demonstrates the better-now-than later value for her. As the value of time goes up, interest rates go up. With higher interest rates, more people see value in lending, but fewer see value in borrowing. Reduced interest rates have the opposite effect. This simple description correctly portrays the dynamics of loans from the points of view of lender and borrower. Interest reflects the value of time and dynamically regulates the amount of borrowing.

From the perspective of the overall economy, interest serves a different but compatible function. To see that, we must return to the marketplace.

Imagine a simple marketplace with many participants who have businesses of equal size. The marketplace expands ten percent per year. Every participant, anticipating continuing success, borrows money at ten percent interest. Suppose that the borrowing exactly equals the total of money in circulation. Further suppose that the government did such a good job last year that it has a holiday this year. It will neither spend nor mint nor lend more money. Finally, suppose that all borrowers succeed in increasing their production ten percent as they had promised.

Everyone is happy at first; wealth has expanded ten percent. However, when it comes time to pay there is a financial crisis! Money has become scarce. People cannot find enough money to pay back their loans. Deflation sets in. Prices go down about nine percent as buyers now find much more to buy for every coin they hold. More than ten percent of the debtors lose the battle over scarce money and declare bankruptcy.

While the economy and everyone in it had done what was promised, the financial structure failed. It had not pumped in a ten percent increase in money so that monetary expansion could faithfully parallel economic expansion.

Deflation is peril to the monetary system. From the perspective of the overall monetary economy, we charge interest on loans to create financial pressure to expand the money supply—to prevent deflation. The interest attached to a loan is a purchase order from the real economy to the money economy requesting more money.

Keynesian economics recognizes the role of interest rates as seen from the perspective of borrower and lender. Raise interest rates and people will tend to borrow less. Lower interest rates and they will borrow more. Such financial manipulation works from time to time, but the practice ignores an economic current running in the opposite direction.

Higher interest rates may discourage investment, but their real purpose is to discourage only unproductive investment while accommodating a surge in production caused by a period of successful investment. This is a big difference. While the economy would charge higher interest when good investment presents a danger of deflation, economists have wanted to raise interest rates when there is a danger of inflation. Is it any wonder that national banks find their inflation-fighting and recession-fighting tasks difficult?

In my imaginary example of a simple but expanding marketplace interest fails to prevent deflation and bankruptcy because my assumptions do not allow it. That is not the real world. A national government that is not on vacation can infuse an economy with needed cash by minting and spending money. However, the public need to spend and the private ability to make productive investments do not necessarily coincide. A national government needs another way to get money directly into the private investment market. Typically, through some kind of central bank, government infuses more money by becoming a lender. In U.S. America the Federal Reserve and Treasury lend to private banks that invest in the expanding economy. Through direct investment and indirect lending the federal government avoids the peril of deflation.

76. National Loans Are as Paper as Money.

Under popular balanced-budget policy the alternative to raising enough money in taxes is to borrow money. Conventional wisdom says that if a national government wishes to spend more than it has taxed, it must borrow. Otherwise it would have to print and spend worthless money. But where is the wisdom in that conventional thinking? When I lend money to the government I get a money-market certificate (or other such paper) in return. My money-market certificate is written on much less expensive paper than my money is. Both are printed on government presses and backed by the full faith and credit of the U.S. American government.

Why is one piece of paper better than the other? If the national economy goes sour and inflation soars, the certificate will be no more valuable to me than money. After all, I'm going to collect my earnings in money. If the money has no value the certificate has no value. Despite these obvious facts, the Federal Reserve System appears responsible and businesslike when it issues paper obligations backed only by paper money, yet congress and president appear irresponsible and unbusinesslike when they issue paper money backed by paper obligations of the Federal Reserve. This makes no sense.

A central bank can be an important issuer of money, but this need not be disguised as helping the national government finance its budget. The expanding marketplace needs money. A central bank, sitting at the apex of a banking system, can provide money when the government expenditures do not match growth in the marketplace. That is a deficit of a different kind. When the productivity of the economy is such that the government does not put enough new money into circulation directly through expenditures there is a monetary deficit. The central bank can give money to local banks who invest it where the marketplace needs it. We can call this gift a loan and administer it accordingly so local banks act prudently. More money is needed. If not provided, the monetary deficit in the marketplace could cripple underlying economic growth. A banking system headed by a central bank can fill that deficit.

Banks specialize in the promises-for-promises part of the marketplace. While many promises can remain in IOUs, bank books, travelers checks, and sundry paper forms, we need real money from time to time. Local banks must be able to cash in on their pyramid of deposits and loans. For that they turn to the central bank (and, if the local bank is going broke, to the insurer of banks). If decisions of local bankers accurately reflect the economy, the central bank can safely help them turn their diverse paper fortunes into simpler fortunes by providing them the money their accounting-sheet assets would merit.

So funny.
That bank must issue
this certain paper obligation for its government
so that its government will not issue
this certain paper obligation for itself.
So certain,
so paper,
so, money!

||

77. *No Surprise, No Revenue*

Borrowing by a national government, like monetary recall, can have merit in limited circumstances. It can extract money from the private economy for public purpose. When a central bank surprisingly offers to raise the rate of interest it will pay to its lenders (citizens, banks, foreign interests), or when it raises the percentage of assets that banks must hold in reserve (not lend) the central bank temporarily extracts money from the private marketplace. Such a surprise can be useful in an emergency. Perhaps the real economy and monetary system are terribly out of whack, or wartime mobilization calls for reduced domestic demand. However, governments, especially in an open society, do not run on surprises. National government cannot sustain itself on revenue from surprise borrowing any more than from surprise taxes because it will run out of ways to surprise us. Even when it does surprise us, the money that surprise borrowing brings into the national treasury is incidental by-product of a more important policy. This by-product is nothing to plan on, nothing to run a government on.

78. *National Bank, Slave to National Wealth*

When a national government borrows money it competes with private folks who wish to borrow. That competition tends to raise interest rates above the rates that economic growth would prescribe. An artificially high interest rate affects the economy. Entrepreneurs may have productive ideas and good projections that show a profit under correct interest rates and a loss under higher rates. They may avoid the risk, take the risk and get lucky, or to take the risk and go broke. In the first instance, the economy loses the new wealth; the product never appears. In the last instance the economy may gain something; the innovation may survive; but the entrepreneur gets sacrificed and his lender suffers a loss.

If I have an excellent idea that at nine percent interest profits me, the economy, and the biosphere, what should I do when a government policy raises the interest rate to ten percent and makes my profitable idea a bad investment? If I do borrow at the artificially high rates I will try to raise the price of goods and services I sell. I may or may not succeed, but I become inflationary pressure. Consistent federal borrowing pushes inflation.

When government borrows it attracts money from people who prefer to invest in government's promise to print enough currency to pay them back with interest rather than in an entrepreneur's promise to market a product that will reward them for buying stock or giving a loan. By trying to cover fantasy budget imbalances through borrowing the national government will make private finances ever more responsive to the central bank and ever less responsive to and promoting of underlying economic growth.

A central bank can play an important role in a monetary economy, but finance should always be subservient to economics. The bank should use its monetary tools to keep money synchronized with wealth.

Part Five:
Community Finance
—topics 79-85—

79. In the Tradition of Delegation

The United States of America delegate power to themselves (a tradition) and to private enterprise (another tradition). After a National Aeronautics and Space Administration took the first USA astronauts to the moon this nation of market-oriented states looked for ways to delegate some space exploration to private companies. Why not also delegate to state and local governments? The Missouri Space Administration and The Cleveland Interplanetary Agency might diversify our approach to space—under national and international guidance. Similarly, state academies of science could take responsibility for much of the scientific investment now directed by our National Academy Of Science.

Such more-local entities would be no more threat to national sovereignty than are state and local health, highway, and police departments, or local school districts. Perhaps we do not consider community-based space and science agencies because they are bad ideas. I suspect, however, we would stop long before considering the merit of such proposals. Even though such proposals follow our tradition of delegation, I suspect that we ignore them because we assume that such large, community-based investments are not affordable. For the time being, they are not.

80. Local Wealth, but No Local Money

While wealth develops at all levels, we tend to concentrate our attention at the national level. Jane Jacobs, in Cities and the Wealth of Nations, lays down a solid argument as to why cities, not nations, spawn economic development. Translated into terms that I like to use, she explains the organization of the city as it develops in economic breadth, depth, and density. The organization occurs not only within the city but within its region; local wealth develops inwardly and outwardly.

Jane Jacobs argues that cities are the sole source of economic development. I disagree on this point, of course. Examples of exceptions abound: cityless gatherers developed wealth by spreading seeds of their favored food plants; neighborhoods and families have produced artists, athletes, scholars, and entrepreneurs out of proportion with their neighbors in the city; E=mc2 is a product of Einstein not Bern, Switzerland; often cities seem to be at least as much products of their regions as vice versa, as in the Silicon Valley area of California; professions develop wealth as their members interchange discoveries, inventions, and techniques at regional national and international conferences; ethnic and religious groups can spawn economic development across political boundaries; national and international efforts have produced space exploration.

I quibble with her on the exclusiveness of the city's role in economic development, but I strongly agree with Jane Jacobs that cities are important, and that they suffer much more than an image handicap relative to nations. They also suffer a serious monetary handicap.

Many people complain about the overburdening emphasis of national government. Ronald Reagan did so when he advocated a new federalism in his campaign for the presidency of the USA in 1980. But there is a major stumbling block in the way of turning more responsibilities over to local governments: money gets minted at the national level. Jane Jacobs sees that in a monetary society this simple fact puts much economic development responsibility at the national level, whether it belongs there or not.

Fifty united states form the federal government of the United States of America. The constitution guarantees certain powers to the states. Many see as a problem the ever greater concentration of function and authority in the national government, but few, I believe, see it as more a financial than a political problem. State and local governments have two fiscal constraints that the federal government does not have. They must budget for real (not fantasy) balances, and they must not mint money. In short, they have wealth but no money.

Our ideal monetary system would keep money production coincident with our wealth production. New money would join wealth where wealth grows. It would get there well-timed to be either inducement or reward. Since much wealth develops locally but only national governments mint money, our monetary system has a built-in tendency to be disjoint. The problem is not as severe as it might be. Local Banks, local corporations, and the characteristics of location itself help attract new money to where wealth grows.

Our ideal monetary system would let money flow to where wealth is in jeopardy. The people of U.S. America have turned to their national government often and asked it to spare little expense to fight war, injustice, technological barriers, economic and natural disasters. Given their fiscal constraints, cities and states can do little to help. Only the federal government has the fiscal irresponsibility to respond. Such crisis-solving power leads to federal political dominance over states and towns. If there is to be a new federalism, states, local communities and their governments must find ways to make money—literally.

81. About Location Currency

What must exist everywhere yet must be everywhere unique?—location.

Location is currency for all communities and governments that preside over geographic areas. Unique sets of privileges and obligations, costs and benefits go with residence in U.S. America, the State of Colorado, the City of Fort Collins. Every place has assets and liabilities given it by geology, geography, weather, history, and ecology. As a community invests and disinvests in the resources that are tied to location, that community's location currency rises and falls in value. Good schools and services tend to increase demand for local property. That demand raises property values and the tax base to pay for schools and services. Pollution and crime do the opposite.

This location-based currency system is inevitable but imperfect. Deeds (to a location) can be traded just as money and stock can. As with money, there are delays and leaks between the outflow of expense and the return of benefit. As with money much of the value reflected in the price of a transaction depends upon larger factors. The value of the deed to one site depends, for example, upon a larger locational currency that reflects value from community investment in facilities and services.

Communities often find themselves in competition with one another. One community promotes its own location currency. This may be good for larger society—driving the quality of local school systems upward to attract industry. It may be bad—driving the enforcement of health and safety regulations downward to attract industry. The wise larger community fosters positive competition and discourages the negative.

Despite its imperfections, location currency is inevitable, powerful, and inalienable. It is inalienable because it is something that no location-based community can give up. Thus, local economic development logically starts and ends with attention to the value people place on living, working, and visiting there—the value of the community's location currency.

82. For Local Banks

Local banks tend to issue money
as stock in local knowledge.
Thousands of local banks
(together and at odds)
foster diverse and robust investment.

Forsake local banks for
  fewer,
  larger,
  national
  and international banks
and you foster
  monotony,
  vulnerability,
  foot in the mud,
  butt in a rut,
  a people and nation
  of downward mobility.

||

Banks expand money supply when they lend out most of the
money they are "holding" for depositors. Thus, local banks
generate a somewhat local currency. Given the lack of
public minting powers at the local level, private local banks
have an important role to play. They have built-in incentives
to invest in local economic development that will return local
deposits. The money that the bank introduces locally to the
national economy responds to—even rewards—success of
local investors. Given a centralized, national monetary
system, decentralized, local, private banking is a
complementary aid to local economic development.

83. For Local Corporations

I like local banks because I believe they promote local
economic development and robust, national economic

diversity. I like local corporations for the same reasons, and one more. While money issued by local banks is national currency whose value comes from the national economy, the stock issued by a local corporation gets much of its value from the value of the corporation that issued it.

In the USA there is waning tradition favoring local banks, however there is little tradition that would restrict corporations to local operation. Now that local businesses buy and sell in a marketplace that is every day more global, such a tradition is unlikely to evolve. Nevertheless, when a community does spawn a local corporation the corporate stock is a share of local wealth (albeit restricted to the assets of the corporation). The value of the stock rises and falls with local corporate value.

To the extent that a corporation participates in and reflects the local economy, its stock is a form of local currency. It is a legitimate medium of exchange whose value is determined by local economic development. By encouraging the creation and permanence of successful local corporations that issue common stock a community indirectly mints local currency.

84. For Local Taxes

State and local governments have one advantage over their national government. In theory and practice they balance budgets with true taxes. National monetary recall is worse than useless. Like trying to pull yourself up by your own boot straps, nothing is gained. When our monetary policy recognizes this we can leave money collection to state, local, and international governments. Absent the illusory burden of national taxes through monetary recall, citizens may approve local taxes to balance their budgets while expanding and improving services.

The value of the dollar in the USA reflects some discount for an underlying local tax burden, but individual local governments do, nevertheless, get more income when they raise taxes. Because they do not mint money, when they tax they collect money minted by someone else. They collect something they need but cannot make, something their constituents have. They collect a type of "tax in kind."

85. For Local Currency

If we wish to pass power and function back to states and localities we may have to give them some of the fiscal irresponsibility that national government has. We may have to let them mint money. This may seem odd, but consider that banks are allowed to mint money when they lend out most of the money that they are supposed to be holding for depositors. Individually we mint a sibling of money when we take out a loan guaranteed only by our promise to pay it back. Corporations mint a fraternal twin of money when they sell common stock. Perhaps it is more odd that state and local government are left out of the minting business. They can issue bonds, of course, but these must be backed with solid guarantees; money is not so backed.

Where I offer fewer answers
may I proffer one more question?
Could your local government
(your community)
mint its own money?
  Miami Marks?
  Saint Paul Dollars?
  ...
I leave it to you
to tell us how to make it work.

‖

The USA has tried "revenue sharing" as a way of channeling money and responsibility back to state and local governments. "Mint sharing" would be the correct approach to such a policy, since the federal government raises no true money-revenue. Local governments could share minting privilege as banks do. The challenge would be to induce each jurisdiction to mint money in proportion to how much wealth it is creating. Absent such inducement, not responsible for the national economy, each local government would always have an incentive to print more money. The European Union faces a challenge as it develops a common currency. Will member nations get no local rewards and incentives for creating the economic growth that allows more currency to be minted? Internationally or intra-nationally, with good carrots-and-sticks to keep local production of currency coincident with local production of wealth, mint sharing could be an exciting evolution in our use of money.

86: In the End

We know.
We know we know the answers.
What to do.
How.
And why.
We know we know.
The false.
The true.
The tried.
Few
are the questions,
and fewer
the answers we cannot grasp.
And at these few
we know to wonder.
Indeed, this is life gifted full:
to know what we know,
yet know how to wonder,
Yet, fuller still in our discovery:
they know,
and they wonder too.

We
and they.
Know
and wonder.
Path to peace!
To human perfection?
No.
This path remains our dream.
And we must know that we know why.
And we must permit ourselves to wonder
how future binds to history.

Today, again:
at which we wonder, they know;
at which we know, they wonder.

At Which We Know, They Wonder.

Plum Local's Appendix:

Plum Local, March 11, 1980

and the

The 2nd Plum Local, November 11, 1981

PLUMILOCAL

Pardon my boldness but I've been yearning to put ideas on paper and share them with others for a long time. A few of these are presented here. There may be another issue of Plum Local in which I can develop these and present some others. In the meantime, I hope these pages of words are of interest to some of you.

1965 <—> 1980

I know.
I know.
I know the answers really easily.
But they're alright.
For it which I cannot grasp.

I wonder.

This is great
    to know
    and to wonder.
It's life really full.
And to make life even
    more perfect.
    she knows and
    he wonders too.

They and I
knowing life
wondering at the unknown--
peace results!

World unites!
Happiness to all!
No these last are just dreams

And I know why.
--Permit me also wonder-
For it which I wonder.
And sit which I know
    they know,
    they wonder.

Fortuitous
The Pulitzer Prize for civilization's most profound work was granted today to someone...

ation of supply, demand and utility, and the elegant description of wealth as the product of energy and knowledge there lies a muddle of witchcraft called macroeconomics.

I found I could not isolate my struggle to the middle of the sandwich because there seemed to be something lurking out there even [beyond?]

than Fullers elegant law. While I have not completed my struggle, I do have several parts of it worth evolving as

My macro economics, I do not know which of these thoughts is unique, is professed by others or discarded by others. I will cite the influences that are obvious to me.

I believe that a valid view of world economics will fulfill certain conditions:
■ It will be possible to derive the laws of market economics from world economics but not vice versa. Conventional macro economics views the world as that

KOMIVESIAN ECONOMICS

PART ONE: IN THE BEGINNING

Since having the crisp logic of market economics brilliantly explained to me by H. James Brown, and simultaneously the crisp logic of Buckminster Fuller's world economics brilliantly set before me in the corner of one of his books, my mind has been struggling to find a path between. Sandwiched between the straight-forward explan-

© 1980 R. Komives

biggest of all markets in which money, economic growth, fiscal stability etc. can be explained in the terms of the market-distribution of scarce resources supply, demand...
The fact that it doesn't work has spawned much debate but little improvement.

■ Economics will be seen as an understanding of one species' increasing independence from its ecology. (Dwai and followers whose outerspace as mans ultimate habitat have developed this fundamental understanding)

■ Wealth will be shown to be the summation of all knowledge (I prefer this formulation to Fuller's Energy + Knowledge)

■ like the biblical loaves and fishes the more wealth is divided the more it multiplies.

■ Economic growth is the inevitable product of cultural evolution (growth in individual and collective knowledge). The challenge is not how to cause it to happen, but rather how to allow it to happen. The valid world economics will show the tie between economic and economic evo-

lution. The current theoretical development of ties between social and biological evolution (sociobiology) will prove to be a derivative of this broader understanding.

Now I will put forth my best outline of World Economics.

With continual interaction of energy forms in the universe, the occurrence of a group of interactions that could purposefully reproduce and multiply themselves by gathering and ordering more energy was inevitable. Life may have occured sometimes on earth or elsewhere

knowledge of how to sustain the reproduction and multiplication of life is stored in the DNA molecule & the genes.

As life multiplied the portion of the universe's total energy physically bound up in life forms grew. An even larger portion was at the disposal of life's energy using processes. The first long lasting life chain needed a system of genes that usually worked but failed often enough so that random but slow natural error could allow adaptive evolution.

→ an inevitable product →

7-DAY WEEK   6-DAY WEEK

The Interplanetary Research Institute is looking for 100 international volunteers, a beautiful and isolated place with accommodations, and a sponsor for project "Eight-Six". One hundred people will spend eight weeks living a daily cycle of 28 hours. The rationale and purposes for the experiment are:
• "Cave" experiments with man's circadian rhythm have shown it to be variable with a greater tendency to be slightly more than 24 hours than to be slightly less.
• The onset of space travel and colonization requires the adaptation of "Space Time".
• The inherent political prejudice in any 24 hour system may interfere with a cooperative-peaceful international endeavor more than the shape of a table interfered with peace in Vietnam.
• Fewer meals and longer waking hours may be physiological better in the low stress space environs.
• 10,000 fewer onstaffs for circadian machinery per century.

179

of evolutional life forms occurred that could store knowledge about purposeful energy capture and utilization (in ways other than genes and DNA) of great survival benefit to these knowledge systems. No lasting ecological balance was ever struck such the fraction of the universe's energy flowing into life and the patterned knowledge systems continually increased. Because the rate of flow was slow relative to the life span of any individual creature, knowledge, patterned behavior tended to be as if the environment were, in principal static. The principal patterns were forms of cooperation and competition that added survival in a world of apparently limited resources.

- BIOLOGICAL KNOWLEDGE is stored into the reproductive code of genes & DNA.
- ENVIRONMENTAL KNOWLEDGE stored from experience in the organs of individual creatures — principally in brain & nervous system.
- ARTIFACTUAL KNOWLEDGE in two forms — utilitarian objects, and coded and preserved messages.

Life forms interacted with one another and with non-biological parts of their environment. Random initially, these interactions became increasingly patterned as they were educated in the three knowledge systems. No lasting ecological balance was ever struck such the fraction of the universe's energy flowing into life and the patterned knowledge systems continually increased. Because the rate of flow was slow relative to the life span of any individual creature, knowledge, patterned behavior tended to be as if the environment were, in principal static. The principal patterns were forms of cooperation and competition that added survival in a world of apparently limited resources.

AGAIN, inevitably, within...

...ENERGY BEGAT MATTER
MATTER BEGAT LIFE
LIFE BEGAT KNOWLEDGE
KNOWLEDGE BEGAT CULTURE
CULTURE BEGAN......

©1980 R. Kephris

one species a pattern of knowledge evolved that could purposefully reproduce itself and multiply. Man is the present if not the first species to be part of such a purposeful pattern. Culture is probably the best term to name this phenomenon in which knowledge takes off and grows and changes faster than the species that has the culture can multiply and evolve.

Just as at some point in time inert matter occurred in a pattern that became life, at some vague point in his evolution man's culture took on a life of its own. Ignorant of any change, men and women continued to compete and cooperate similarly to individuals in other species. However, because of the cultural DNA and cultural genes that were at work, their child was likely to have greater knowledge than they had possessed at the same age.

The key group of cultural genes has been that responsible for reproduction of government. These genes have been persistent so that as in many species, each individual is the number of one or more groups that have a system for governing itself and multiply. The genes are also more bound, so that, unlike other species the form of government undergoes change. The changes having the most success at investing knowledge with a high rate of return in more knowledge have the longest time of survival.

Government invests by allocating group knowledge in ways individuals cannot or wish not to allocate it. It can, for example, create a system of attack and defense that is more effective and occupies freer people than with everyone depending on himself. It can guarantee someone who wishes to experiment with a new technique for hunting that he or she will have something to eat if it fails. It can set up and enforce rights and privileges that will give some order under which individuals can cooperate with minimum risk.

Must of cultures knowledge is directly or indirectly tied to these communal investments made by group governments. This evolution of civilizations have virtually disappeared over night when a government structure collapsed into the hands of people who could not adequately maintain and propagate the communal knowledge. Wealth disappeared as hundreds of civilizations flourished and withered.

In other civilizations the rate of knowledge expansion slowed down as very stable governing found stability in commanding the growth and distribution of knowledge. Across the breadth of the human species however group investment kept the general cultural advance in motion.

One variant of man's governing propensity-genes was the creation of instant and ephemeral governments for the purpose of trading knowledge with people outside their group. As an alternative to making one's private goods and exchange, privileges with other peoples provided some mutual rules could be agreed to that would...

MINIMIZE the risk.

Vast redistribution of knowledge took place continually as such trade became common. The instant government of mutual agreements that allowed such trade were not always done centrally by some central metal as some continued for thousands of years.

When cultural growth began to happen so fast that the communal distribution of wealth through group government had difficulty distributing and investing new forms — factual knowledge of — the trading became a way to distribute goods within the group. Bound only by the governing rules of the market place each trader can use his eloquence and competitive instincts to exchange with others.

Just Use Vote Only
James Carter = Yes
James Carter = No
R. Reagan = Yes
R. Reagan = No
☐ ☐ ☐ ☒
Could this form be substituted?

©1980 R. Kephris

Incorrect "Urban Solar" Architecture... derided by glaring angle

Better Urban Solar Architecture... derided by solar waves
... derided by solar waves
proper coverage for ef-
ficient exposure for ef-
ficient on an optimum angle

KOMIVESIAN ECONOMICS

PART TWO: "TAXES, BAH!"

A new world economics has to be of some practical use. I have thought of several. I hope there are many more. I will briefly explore one.

Nothing seems to prod more intense economic interest these days than strategies to balance the U.S. federal budget. These strategies convert to "how can the U.S. government spend no more than it takes in through taxes?" This seems to be a two-sided equation:

$$SPEND = INCOME$$

©1980 E. Komives

EXCLUSIVE REPORT ON THE MINNEHAHA CONCLAVE
by Chris Kurt

As the 1980 elections approach I can't believe that I am the first to break the seal of secrecy over the events of Halloween, 1978— The last days before the last national election. I hope that by revealing what happened that night I can save others the tragic disillusionment from which I am now almost recovered.

There were so many conservative Republican candidates, conservative Democratic candidates, an almost-tread-on-me independent candidates trying to get elected and dismantle the government that year that I decided to seize the opportunity. I awoke in a sweat. I was trying to get over a nightmare about having a secure government job to turn to if all else failed. It was then that I decided to put out a call to a secret conclave.

I gathered some brief but detailed instructions, telegraphed them to the politically astute friends we have left in the political wake of our peregrinations across the country, and took off to make preparations at the conclave site. Taking a shut from the College of Cardinals, I ordered up a fan of Ritz Crackers and a thousand bottles of mineral water. These would serve as our only sustenance until the conclave had done its thing.

Then it began. While kids across the country were home getting sick over candy they had been collecting all evening, while voters who had taken one night off from eagerly attending political rallies were relishing the tricks they had played on the little pests in their neighborhood, while all this was happening, the disguised candidates began arriving in procession to my conclave — each with one lighted candle in one hand and an attaché case filled with extras...

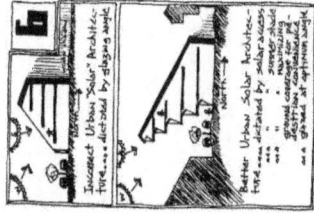

Swamp calculating this figure is just as important as estimating which wealth investments will be productive in expansion of knowledge.

Market economics is an effect in as much a cause of man's growth in wealth. It always has and always will be adjusted by communal decisions when it gets distorted relative to world economics. War and political collapse have been frequently used to correct distortion. The "New Deal" and World War Two slowed massive communal investment of knowledge where the market indications of the great depression were that there was little wealth to invest. The humanistic goal of an enlightened world economics is that such trauma can be removed from the wealth distribution and expansion. Trauma has been a necessary part of man's unconscious role in cultural expansion. As a conscious participant he will be able to invest profitably and more efficiently in the peaceful expansion of knowledge.

End of Part I
©1980 R. Komives

Macro-economics is man's conscious involvement in this inevitable process. By understanding the principles of knowledge formation as the principles of economic growth, and by guiding our group government with that understanding we can decrease the probability that man's culture and the human species are on an evolutionary track leading to a dead end.

Most wealth is undivisible after a variety of communal reasons, not in the market place. A heating stove is found in the market place. It is a storehouse of functional and possibly aesthetic knowledge. Yet that wealth, in the form of pleasant or even life-saving warmth is likely to be distributed thru communal tradition or law to many passers by who give nothing in trade.

Thru these processes culture has reproduced itself and expanded. Man was the conscious vehicle, but not the conscious cause. Nor nature could destroy the culture or even the life farms. However, the process of evolution from matter to life thru knowledge to culture would begin again.

The basic formula:

$$W = K_h \cdot K_b \cdot K_e \cdot K_a$$

Where K_h = human wealth
K_b = Biological knowledge
K_e = environmental knowledge
K_a = artificial knowledge

We are probably less than one millionth expresses the fraction of total energy in the universe that is either physically incorporated in the knowledge forms or being used by it. A solar collector and the energy being captured by it, plus a refrigerator and the food within it, plus the brain cells storing the learned ability to swim and the water supporting the

Anti-government candidates are to-a-man and to-a-woman practical. Thus it was, that by 11:53 p.m. central standard time, in the freezing cold confines of the Narrow Canyon formed by Minnehaha Falls in Minneapolis, 315 conclavers were assembled. Like chair boys in down parkas and Lone Ranger-type Halloween masks they were ready to go to work. I took 27 minutes for my opening remarks. In sum, I told them of the inevitable victory of their cause. It would soon be within the power of this esteemed group to create a utopian country that had no government or bureaucrats. The people of the country were speaking. These 315 men and women there assembled were the best listeners, duly two necessary elements for the ultimate victory were missing.

One missing element was their own combined conviction that the holy mission on which each had individually embarked could be accomplished if the conclavers were willing to march as one in a crusade. "Are you ready for the crusade?" I whispered emphatically. My question was answered by four minutes of delirious shouting, embracing, hand-shaking and the chant "Hitler again, Hitler again harder! harder!"

After joyfully raising my arms to restore calm, I told them that the second element missing was a simple program for implementation. It could be a simple program because all that is necessary is to ease the bureaucrats out of their offices and let private enterprise bid on the vacant facilities. A few other details in the program needed to be worked out, but it could all be planned by sunrise—in time for everyone to get back to the campaign. After all, the program wasn't so terribly necessary in itself. We would just have to be ready with good information and PR to head off the self-serving attacks that would be organized by the old guard and the socialists.

It was now midnight. The conclave set about its task. Each of twelve sub-conclaves, whose members were chosen by lot, was to produce one

7

Thinking of income in terms of the economics of knowledge and wealth, government has historically been inconsistent yet phenomenally successful in having income exceed expenditures.

SPEND <<< INCOME

This is true in the US over the last century even though the popular image is of a country outspending its resources. Part of the problem is that taxes have been used as the measure of a large part of that income.

What are taxes? They are a reduction in the percentage of market wealth that can be invested by individuals. Every dollar in someone's hands is convertible in the marketplace to a small percentage of the artifactual and environmental wealth that is being traded (in times of slavery biological wealth could also be found in the market).

The government is supposed to use the market shares it has gotten thru taxation to make expenditures. There is a serious problem with this concept however because

paragraph for our program and proclamation.

I passed out Ritz crackers and fielded several questions from the various sub-conclaves. ......

"Hey, our group decided to keep our armed forces and the police. Otherwise, anarchists and communists would run roughshod over us. Is that alright?"

"Sure but figure out how to administer them without bureaucrats."

"We're not sure whether Greyhound, Brinks or General Motors is capable of taking over the street and highway system. Does anyone have any ideas?......O.K. we'll keep working on it."

"Some of us westerners were wondering if it would be possible to rewrite some of the history books our kids are forced to read in school. That bit about the government dividing up land, giving it to our grandfathers and building water projects, and subsidizing our production doesn't sound too good for our cause."

"Don't worry. Group 7 is doing away with the schools. You'll be able to keep your kids at home and teach them what you want."

"We're a little worried about the possibilities of revolution if we eliminate all of the health, welfare and civil rights programs at once. Do you think it's alright to increase the army a little to keep everyone in check, or to phase programs out slowly so no demand you come along and incite riots?"

"Hmmm, well I'm afraid that if we did the latter our own supporters would call us the wishy washy conclave rather than the Minnehaha conclave."

"Hey, if anyone objects to having a cell in the basement of each home to handle convicted criminals on a rotating basis let us know. That's the way we are thinking."

The Ritz crackers were consumed by two o'clock. The rumble-rumble-rumble questions ceased by 2:30. The

8

taxation itself has costs. No matter how carefully one tries to pour from one bucket to another spillage occurs. If a government truly wishes to spend as much money as it took out of the hands of its citizens it has to print a little money to make up the difference and hope no one notices.

What happens when the government prints up money and spends it?
1. Any government expenditure is an investment of its citizens' communal wealth in items which are by choice or by impossibility not to be private investment.
2. More important to this discussion, expenditure of nonexistent money is a reduction in the percentage of market wealth that can be invested by individuals. Each new dollar reduces the market share of the old ones.

Taxation does one thing. Spending does two—one of which is identical to taxation. This, the fundamental problem with the concept of balancing spending with taxation is they are both all the same side

© 1980 R. Konikow

of their voices put me to sleep. I awoke with the first rays of sunrise to find the whole conclave standing over me. They looked tired. They also looked a little discouraged. Yet many seemed ready to burst into a smile.

Then one of my southern cowlackers drawled, "Heah it is." I was puzzled when the piece of paper he gave me in the dim light was clearly a page torn from a book. He and each of the other conclavers patted me on the back as they filed up and out of the little canyon. Each said something to me like, "We did it." or "That's just the first dozen lines, but you can fill in the rest." I began to get the idea.

My suspicions were confirmed when I glanced down at the page I was given. The light was now just bright enough for me to read the small printed words, "We the people of the United States, in order to form a more perfect union, establish justice........."

The next thing I remember is my wife reviving me in a pleasant cell at a Minneapolis police sta- tion. They had called her in from Colorado. This Halloween goblin who appeared to be her husband had been found skipping merrily thru Minnehaha Park singing, "Kinky commie capouts! Kinky commie Capouts! Kinky commie Capouts!......."

PLUM LOCAL

PLUM LOCAL has served its principal purpose just by getting printed. Just in case the words have some value I retain all rights of copying and reproduction in any form. However, I would be thrilled if anyone wanted to quote... In this spirit, and in recognition of the rapid ways in which anyone can distribute printed knowledge today, I gladly authorize the copying or recopying of PLUM LOCAL #2/100 provided the following conditions are met and two maps must be rapid and distributed together, copy

2 copies can be given away free of charge but may not be sold, (or needed) 4 the copy given each batch made, together with 10¢ per each copy ready, must be sent to me. soon after the copies are made.

This will feed my ego and also locate experiment in anarchic ways to approach freedom. I pay each... individual author distributic reproduction and storage. That is I... would think, I'll tackle that wealth growth another than most... opportunity next time. In the meantime wonder is under pressure to stop or recopy goodtimes. If you have read this far, I'm pleased —Robert Koming, 321 East Plum St, Fort Collins, Colorado, 80521

9

of the egalitarian spending is taxa- tion. Taxation to support government expenditures is unnecessary. It is also counter productive, be- cause it consumes wealth in a useless circular bucket pouring exercise, and because it is inflationary. (since everyone can estimate the approximate share of dollar income that will return to the govern ment he or she will act in the market place as if the money is worth less than face value)

Taxation is useful for only one purpose when imposed by the same level of gov- ernment that prints and coins money. It can be used to distort the mar- ket—either by changing relative price (beer vs root beer), or by removing some curren- cy from circulation. The latter only works if impos- ed rapidly and for one- time-only collection. Other wise, the market inflates itself to compensate. Money raised for either purpose or distortion can be destroyed or recycled, but can never be used to legitimize a budget or an expen- diture.

For levels of government

---

that do not print at mint money—family, United Na- tions, city, Boy Scouts, etc.— taxation does have another purpose. When support is either not desired or not received from the minting government, other levels of communal groups can tax their market resources to enable communal actions in the market place. This is legi- timate budgetary income.

National expenditures always have and always will be balanc- ed only by growth in wealth caused by the communal and private reinvestment of knowledge. Inflation may indi- cate past investment; it may also indicate a divergence of the marketplace from real wealth formation pa- tterns. It never indicates that there has been too little tax income to cover investments.

"Taxes, Bah!" Let's phase them out over the next twenty years, except for the limited purposes for which they can be of some use. Let's balance our budget by investing communally (politically) in the growth of knowledge for mankind.

END

PREDICTION: ☑
The typical US child born in 1980 will grow up not knowing what NBC, CBS, or ABC are.

This second issue of PLUM Local has been a longer time a-brewing than I had hoped. There have been enough kind inquiries to spur my productivity. So here goes...

1480 ⟷ 1981

Hanging upside down
In a world of frowns
Is the only way to see a
smile these days.
Hanging upside down
turns the sky to ground,
And pebbled clouds
show the way.

People frown &
say you're so
crazy now;
They see a clown
just hanging
in jest.
They don't see
the downside
up benefits
of seeing the
world at
its best.

music by: K. Korb

Lyrics by: K. Korb &
B. Komives

Fort Collins, CO.

The Brain: The Last Frontier; by Richard M. Restak (Warner Books) has been nominated by the Unfortunate Friends of PLUM Local for the Plumitzer Awards:
Excessive Enthusiastic and Frequent Citation;
For its ability to pop up in any conversation on any topic if Komives were in the vicinity.

KOMIVESIAN ECONOMICS

In PLUM Local I an alternate view of world economics was described as an explanation of increasing wealth in the biosphere and for the human species. This economics of abundance contrasts with conventional economics of scarcity. In "Part Two" the discussion jumped to a specific disagreement with conventional views of taxation and money.

It was argued that taxation carried out by national governments is worse than worthless. More specifically:

- Any regularly imposed tax by a government that also prints or mints money generates no income for that government; rather

- It inflates the currency (decreasing its value); and

- It's supposed task (to generate income) is efficiently carried out by the printing and spending of money.

Now I wish to share a related insight that both supports this view of taxation and money and relates it to fundamentals of the biosphere...

THE ROLES' ROLE

"Role playing" & "Role differentiation", terms of frequent usage in sociology, are also apt titles for a mechanism used in the biosphere to protect and expand wealth. They are also better and more general terms for an economic concept called "taxation."

Whenever in the dawn of life two separate organisms first accidentally contributed to eachothers chances for survival and also produced offspring that had a tendency to be similarly supportive, role playing became a part of the system of knowledge that supports and increases the amount of life in the universe.

Each particle of DNA, each organ, each species, each individual, each group depends upon and is depended upon by other elements of the biosphere. Despite some apparent major traumas in prehistory that threatened the continuation of life, the % of the universe's energy that was incorporated into life and life support functions has been increasing. The engrained knowledge of role differentiation has been both a necessary cause and a changing product of this flourishing of life. By role playing the elements of the biosphere cannot only prevent (absent any ultimate trauma) the loss of energy from biological matter, but can actually cause a net energy increase. In biosphere economics, wealth has been steadily increasing.

In human economics, the same phenomena are present. The science of ecology tells us how we share the wealth of other species for our benefit. Similarly, within our species, there are no true individuals.

*net loss

Bi-sexual reproduction & infants who are dependent on adults for survival require, at a minimum, that we develop family roles. Unlike other species, homo sapiens, have retained a substantial amount of flexibility in the structure of these roles. Matrilineal, patrilineal, matriarchy, patriarchy, extended, nuclear, monogamous, bigamous, etc. – all descriptions of role playing forms that have proved useful in the history of mankind. In large populations the family role playing has been extended to an undefined point where we call the structure government rather than family.

Role playing works because it transfers the resources among individuals towards the benefit of the group. The infant needs its family to supply it with the ener-

gy needed to survive. Given survival, the infant has the potential to contribute to the welfare of the whole group, and reproduce another generation of the species. This nurturing tendency has been transferred to the creation of scholars, artists, scientists and administrators as well as the preservation of life and opportunities for less fortunate individuals in a culture. Food, education & security are harvested, organized, created and distributed through a system of role differentiation. While history is too full of examples of individuals and groups who were forced into the roles of the exploited and persecuted, the total and average per-capita wealth has tended to increase. The exploitation and persecution are sad products of an ignorance of role playing.

Role playing (1) distributes wealth continually among individuals, and (2) creates wealth in the process, provided the roles are mutually beneficial. Attribute number 1 is well understood by the exploiters & by the other species on our planet. Attribute number 2 is the frontier of our understanding of world economics.

Taxation, likewise, (1) distributes wealth among individuals, and (2) creates wealth in the process, provided it works to the benefit of the larger group. Taxation is role playing.

2

PROPOSAL: FUND THE MY. MISSILE* SYSTEM

To: Ronald Reagan
From: Otis Kom
Subject: Supply side defense

As a productive alternative to thousands of people & billions of $ being buried in useless tracks and deadly darts in the Great American Desert, let's put tens of thousands of USA & USSR citizens in hundreds of mobile bicultural mobile university classrooms, circulating randomly among one anothers strategic targets. This project will cost less, create profit, and be a greater deterrant than the MX.

* "Missile": from Latin *missilis*, to send.

†Legitimate forms of taxation can be classified into at least 4 types. Each type offers certain advantages, but otherwise they are interchangeable ways of fostering desired role differentiation.

- Taxation by law or custom — in which beneficial role differentiation is fostered by a system of rules that tend to govern behavior.

- Monetary taxation — in which government offers negotiable currency to induce people to play the role of providing goods and services in support of public efforts. In short government prints money and buys what it wants.

- Taxation in kind — in which goods and services are provided directly to government. (Shall governments that do not print their own money may legitimately accept money as in-kind taxes)

- Taxation in prices — in which government artificially raises (or lowers) the prices of some goods and services in the marketplace to foster a more beneficial private investment pattern.

There is one very common form of illegitimate or pseudo taxation:

- Monetary Return — in which →

government recalls from its citizenry a portion of the money that it previously put into circulation.

Any law accomplishes taxation. "Thou shalt be available for jury duty," "Thou shalt put x% butterfat in your ice cream." "Thou shalt send thy children to school." "Thy employees shall not work on the 4th of July." Each law taxes the citizenry by inducing individuals to allocate resources to carry out the mandates.

The nationally desired role differentiation that occurs on July 4th is taxation by law. It is also taxation in kind since employers must put up the costs of a paid holiday. Congress could have decided to buy the day by giving employers cash compensation (monetary taxation). It could have required double pay for work on that day. This taxation in prices is actually in force, encouraging industry to stay closed that day. Under any alternative the presumably beneficial role playing is fostered, and the allocation of national wealth is adjusted.

Monetary return such as we have in the national →

income tax system has no role to play. 【3】 If it were necessary to re-collect money before buying the 4th of July holiday from employers, then it should also be necessary to collect a like amount each time we enforce a law which accomplishes the same end. Of course we do not. We need not.

If passing a law can accomplish nearly the same results as printing money, it is also subject to the same dangers of inflation. If the 4th of July holiday is an unproductive idea, it will be reflected in increased prices (demanding more money in circulation). If it is a productive idea, the resources temporarily allocated to it will be overwhelmed by the resultant increase in collective wealth.

I was taught that money is a medium of exchange. Once in the marketplace it does serve that function. However, when first spent by government it's not a substitute for goods and services, but rather a very mobile law. Wherever it is spent, it influences the roles played by individuals in our society. True taxation occurs when money is spent, just as when a law is passed, a traditional family role is played out, or energy is shared by micro-organisms.

Taxation is our tendency to cooperate towards increased Abundance. It is the role of roles.

Blessed Curses

Mankind seems to have two significant attributes that distinguish it from other species, and that must contribute to the unique role of humanity in the biosphere.

We have brains that have two somewhat independent hemispheres, enabling us to process information in at least two different ways simultaneously.

We have social systems that tend towards beneficial stability, but that are frequently dominated by inevitable instability.

We seem cursed by the anguish of war between the conscious and subconscious or logical and intuitive factions in our brains.

We seem cursed by the bloody and bloodless wars between factions in our society.

Other species have evolved generally stable systems of social order. If their nervous system has two-part brains, the parts seem in harmony compared to ours.

Our unstable society may well be a curse brought on by our unstable minds.

well be a curse brought on by our unstable

Our creative ability to explore the workings of the universe thru art, science, engineering and fantasy are probably the blessings we owe to our cursed instability.

If there is a utopia it will be found in a humanistic management of instability.

All species see the instant scarcity of available resources and an apparent need to stifle the redistribution that instability may bring.

Only humans can hope to see the increasing wealth of the biosphere that instability has wrought.

Only humans can hope to see that beneath the rules of scarcity and competition lie the more general rules of abundance and cooperation.

Only humans have the blind power, at this tick in the evolutionary clock of the universe, to destroy life and all change it blesses us with...

Forward to Basics

Index

www.ingramcontent.com/pod-product-compliance
Lightning Source LLC
Chambersburg PA
CBHW022037190326
41520CB00008B/615